Advances in Intelligent Systems and Computing

203

Editor-in-Chief

Prof. Janusz Kacprzyk
Systems Research Institute
Polish Academy of Sciences
ul. Newelska 6
01-447 Warsaw
Poland
E-mail: kacprzyk@ibspan.waw.pl

For further volumes:
http://www.springer.com/series/11156

Srikanta Patnaik, Piyu Tripathy, and Sagar Naik (Eds.)

New Paradigms in Internet Computing

 Springer

Editors

Dr. Srikanta Patnaik
Computer Science and Engineering
ITER
SOA University
Bhubaneswar
India

Dr. Sagar Naik
Dept. of Electrical and Computer Engineering
University of Waterloo
Waterloo
Canada

Dr. Piyu Tripathy
TempleCity Institute of Technology
and Engineering (TITE)
Khurda
India

ISSN 2194-5357
ISBN 978-3-642-35460-1
DOI 10.1007/978-3-642-35461-8
Springer Heidelberg New York Dordrecht London

e-ISSN 2194-5365
ISBN 978-3-642-35461-8 (eBook)

Library of Congress Control Number: 2012953493

Printed on acid-free paper

Springer is part of Springer Science+Business Media (www.springer.com)

Preface

The internet has changed the way we use computers, and it is still changing. This programme will develop expertise in key technologies and shape the internet computing. Internet computing uses the internet and central remote servers to maintain data and applications. It allows consumers and businesses to use applications without installation and access their personal files at any computer with internet access. This new paradigm in internet computing allows much more efficient computing by centralizing data storage, processing and bandwidth. This volume includes seven chapters of internet computing from various research and academic instuitutions.

First Chapter entitled "Enhanced Mobile Ad hoc Routing Protocol using Cross layer Design in Wireless Sensor Networks" by Waris Chanei & Sakuna Charoenpanyasak elaborates the deployment of wireless sensor networks consist of several tiny sensor nodes for monitoring the environment conditions. They have used IEEE 802.1.5.4 standard to forward the data from each sensor node to the base station wirelessly. Due to the limited resources in wireless sensor networks, they have proposed a routing protocol to consider the energy level. In their report, they have proposed a new method based on AODV routing protocol to find the route using cross layer design, which can save the energy and reduce overhead compared to the original AODV.

The second Chapter entitled "Decision support in supply chain management for disaster relief in Somalia" by E. van Wyk, V.S.S. Yadavalli and H. Carstens addressed some of the issues in supply chain management with the trade-off between stockpile cost and shortage cost by using pre-emptive multi-objective programming in Somalia. The inadequate infrastructure and poorly planned logistics of Somalia may lead to the destruction of the country. To address these concerns, it is necessary that humanitarian aid is pre-positioned to provide victims with sufficient relief. This chapter proposed a model followed by a case study based on Somalia, illustrating the functionality supply chain management with the trade-off between stockpile cost and shortage cost.

Chapter Three entitled "Challenges in Cloud environment" by Himanshu Shah, Paresh Wankhede and Anup Borkar is a white paper on cloud computing. They have addressed a real life situation of the IT industry. They have quoted that Organizations have been skeptical about moving from traditional data-centre onto cloud. Along with well known concerns like security, loss of control over infrastructure, data theft, lack of standards, etc; cloud does pose issues like portability, reliability, maintainability, ease-of-use, and etcetera. Their paper discussed about these concerns around system quality attributes using Amazon Web Services (AWS) and Azure cloud as reference. Their paper encompasses the recent challenges faced and probable solutions for the same and suggested some management tools based on the parameters related to system quality attributes such as portability, reliability, maintainability, etc.

The authors have also presented another paper entitled "Challenges on Amazon Cloud in Load Balancing, Performance Testing and Upgrading" on web application hosting in a data centre. They have discussed several issues ranging from hardware provisioning, software installation and maintaining the servers and also timing and cost aspect of the procuring and software delivery. They have highlighted the pitfalls encountered and possible resolutions for each and endeavors to find out the best architecture which would give maximum return on investment.

Chapter Five entitled "A Rule-based Approach for Effective Resource Provisioning in Hybrid Cloud Environment" by Rajkamal kaur Grewal and Pushpendra Kumar Pateriya elaborates the important issue of resource provisioning in cloud computing and in the environment of heterogeneous clouds. They have propose a *Rule Based Resource Manager* for the Hybrid environment, which increase the scalability of private cloud on-demand and reduce the cost.

Chapter Six entitled "Browser Based IDE to Code in the Cloud' by Lakshmi M. Gadhikar and et. al. have proposed a cloud based Integrated Development Environment (IDE) for the Java language which will have the features to write, compile, run and test code on the cloud. They have designed a software, which allows easy development of Java applications and also provides sharing of code and real time collaboration with peers. This can be used by the developers who require instant help related to coding Java applications. They have also mentioned that the IDE also eliminates the need to use conventional devices like desktops or laptops to code programs by allowing the users to access this from various devices like smart phones with an Internet connection.

The last chapter but not the least entitled "Zero Knowledge Password Authentication Protocol" by Nivedita Datta has proposed a simple protocol based on zero knowledge proof by which the user can prove to the authentication server that he/she has the password without having to send the password to the server as either cleartext or in encrypted format. Thus the user can authenticate himself/ herself without having to actually reveal the password to the server.

Dr. Srikanta Patnaik
Dr. Piyu Tripathy
Dr. Sagar Naik

Contents

Enhanced Mobile Ad hoc Routing Protocol Using Cross Layer Design in Wireless Sensor Networks

Waris Chanei and Sakuna Charoenpanyasak

Center of Excellent in Wireless Sensor Networks
Department of Computer Engineering,
Faculty of Engineering
Prince of Songkla University
Hatyai, Songkhla, Thailand
c.waris@hotmail.com, jsakuna@coe.psu.ac.th

Abstract. Wireless sensor networks consist of several tiny sensor nodes largely which were largely deployed to monitor the environment conditions. Each sensor node will forward the data wirelessly to the base station using IEEE 802.1.5.4 standard. The wireless network management will be different depending on the applications. In wireless sensor networks routing protocols, there are many control messages. Because of the limited resources in wireless sensor networks, routing protocol has to design to consider the energy. This paper proposes the new method based on AODV routing protocol to find the route using cross layer design. The proposed routing protocol can save the energy and reduce overhead compared to the original AODV.

Keywords: Wireless sensor networks, Routing Protocol, Mobile Ad hoc Networks, AODV, Cross layer design.

1 Introduction

Wireless sensor networks (WSNs) have witnessed a growing interest in deploying. Sensors are expected to be low cost and can be applied to many applications. In the network having many sensors using multi-hop communication is required the routing protocol for to find the communication path and maintenance the route [1]. There are many routing protocols in propose for WSNs depending on their applications. In this research focus on multi-hop communication, therefore choose AODV routing protocol. But AODV designed for mobile ad-hoc network (MANETs). This protocol is not appropriate for WSNs because of there are still many limits of energy and resource [2]. Sensor node to send – receives packet low remaining energy than other sensor node. AODV protocol use route discovery to find a new path. But in route discovery is not energy parameter. If AODV choose low energy path. When some node in path is power off. AODV broadcast message control to find a new route. It effect to high routing overhead, waste energy and bandwidth.

This paper proposes the new algorithm to improve AODV for WSNs applications. Modified AODV the route discovery called Energy Weight –AODV (EW-AODV)

S. Patnaik et al. (Eds.): *New Paradigms in Internet Computing*, AISC 203, pp. 1–11.
DOI: 10.1007/978-3-642-35461-8_1   © Springer-Verlag Berlin Heidelberg 2013

has used the cross layer design technique for routing layer   to access the energy parameter from the lower layer. The Network Simulation 2 (NS-2) is employed to analyst the results between EW-AODV and AODV.

This article describes the operation of wireless sensor networks on IEEE 802.15.4 standards, AODV routing protocol and Cross layer design method in Section 2. The EW-AODV describes in Section 3. The simulation parameters and network topology using NS-2 network are presented in Section 4. The performance analysis and comparison are present in Section 5. The conclusion is described in Section 6.

2   Theory

2.1   IEEE 802.15.4 Standard

IEEE 802.15.4 [3] is a standard which specifies the physical layer and media access control (MAC) for low-rate wireless personal area networks (LR-WPANs). The basic framework conceives a 10-meter communication range with a transfer rate of 250 kbps. Trade-offs is possible to favor more radically the embedded devices with the low power requirements, through the definition of not one, but several physical layers. The transfer rates of 20 and 40 kbps were initially defined, with the 100 kbps rate being added in the current revision. This standard supports topology one-hop star and multi-hop with the frequency band at 2.4 GHz or 896/915 MHz size of payload 104 bytes and address length 64 bit or 16 bit (support 65,000 nodes. The sensor can communicate using multi-hop and also required a routing protocol to manage the path and send the information.

2.2   Ad hoc On-Demand Distance Vector (AODV) Protocol

AODV [4] Routing is a routing protocol for mobile ad hoc networks (MANETs). AODV is a reactive routing protocol, meaning that it will establish a route to a destination only on demand. AODV find a routing path independently of the usage of the paths. At that point, node has to broadcast a request for a connection. The advantage of AODV is that there is no overhead traffic for the communication along the existing links. Also, the distance vector routing is simple, and doesn't require much memory or calculation. However, AODV still requires more time to establish a connection, and the overhead of the initial communication to establish a route is heavier than some other approaches. Fig. 1 show AODV Route Discovery, Node 0 is source node and Node 7 is a destination node. Nodes 0 start to broadcast a route request packet (RREQ) to the neighbors node, When the neighbors receive RREQ, it will check in the route table. If there is no route to a destination, AODV will broadcast to other neighbor nodes. When the destination node receives RREQ, it will send a route reply (RREP) to the source node. As we can be seen, the AODV protocols produce a large amount of control message when the route discovery is activated. Therefore, the energy is consumed. In this research, the energy saving routing protocol using cross layer design is proposed.

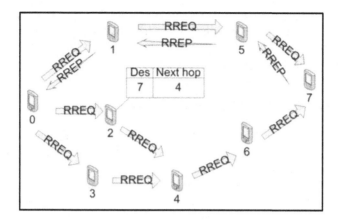

Fig. 1. AODV Route Discovery

2.3   Cross Layer Design

By using cross layer design technique, each layer is allowed to exchange the data across layer [5]. Cross layer can solve some problems and improve the performance of WSNs. Cross layer allows a protocol to share and access the information from the different layer. Information exchange is not required in the adjacent layers for example; MAC layer can exchange information between Transport layer. The communication layer interfacing is shown in Fig. 2 Figure 2A shows a new interface call "Upward" information Flow by creating a tunnel forwarding the information from lower layer to the upper layer. In contrast, the "Downward" information flow as show in Figure 2B is created to allow the lower layer to gain the data from the upper layer. For example the routing layer can set some parameter in mac layer for data transmission. Figure 2C shows the Back and Forth information flow link between two

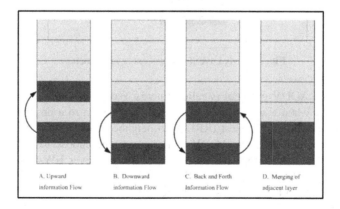

Fig. 2. Cross Layer design

or more of non-adjacent layers. A combination of the adjacent layer can be merged as show in Figure 2D This paper uses the upward information flow between the network layer and lower layer for the energy information. The energy parameter has been used in route discovery and will be described in the next section.

3   Propose Energy Weigh – AODV Protocol

The routing protocol for WSNs should be low complexity and use less resource. AODV is routing protocol designed for MANETs. AODV is reactive protocol, it has lower routing overhead than proactive protocol. Thus AODV suitable for WSNs. Although AODV has low routing overhead, the route discovery is still wasted a large amount of energy due to its control message. So that, the routing protocol is designed to minimize the broadcasting overhead [6]. In WSNs, energy is the most importance. Unfortunately, AODV has no energy parameters. Cross layer design technique can allow the higher layer to access the data from the lower layer. In this research AODV routing protocol can access the energy parameter from the lower layer as show in Fig. 3.

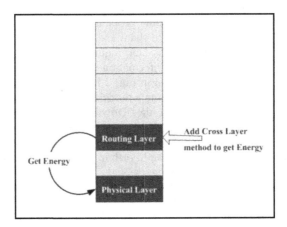

Fig. 3. Cross Layer to access energy parameter

The modified AODV protocol is called Energy Weight- AODV (EW-AODV). In original AODV, when a source node sends the information to a destination node, it has will broadcast Route Request (RREQ) to neighbor nodes. When the destination node receives multiple RREQs. AODV will compare and choose RREQ with higher sequence number or less hop count to send RREP. However EW-AODV adds the energy field in RREQ as can be seen in Fig. 4.

The energy field in RREQ will show the remaining energy of each node in the path using our algorithm RREQ Energy use for get lower current energy node in path with RREQ passes to Destination. It have algorithm to add RREQ Energy value to RREQ as show in Fig. 5.

```
0                   1                   2                   3
0 1 2 3 4 5 6 7 8 9 0 1 2 3 4 5 6 7 8 9 0 1 2 3 4 5 6 7 8 9 0 1
+-+-+-+-+-+-+-+-+-+-+-+-+-+-+-+-+-+-+-+-+-+-+-+-+-+-+-+-+-+-+-+-+
|     Type      |J|R|G|D|U|     Reserved      |    Hop Count    |
+-+-+-+-+-+-+-+-+-+-+-+-+-+-+-+-+-+-+-+-+-+-+-+-+-+-+-+-+-+-+-+-+
|                            RREQ ID                            |
+-+-+-+-+-+-+-+-+-+-+-+-+-+-+-+-+-+-+-+-+-+-+-+-+-+-+-+-+-+-+-+-+
|                      Destination IP Address                  |
+-+-+-+-+-+-+-+-+-+-+-+-+-+-+-+-+-+-+-+-+-+-+-+-+-+-+-+-+-+-+-+-+
|                  Destination Sequence Number                 |
+-+-+-+-+-+-+-+-+-+-+-+-+-+-+-+-+-+-+-+-+-+-+-+-+-+-+-+-+-+-+-+-+
|                       Originator IP Address                  |
+-+-+-+-+-+-+-+-+-+-+-+-+-+-+-+-+-+-+-+-+-+-+-+-+-+-+-+-+-+-+-+-+
|                   Originator Sequence Number                 |
+-+-+-+-+-+-+-+-+-+-+-+-+-+-+-+-+-+-+-+-+-+-+-+-+-+-+-+-+-+-+-+-+
|                          RREQ Energy                         |
+-+-+-+-+-+-+-+-+-+-+-+-+-+-+-+-+-+-+-+-+-+-+-+-+-+-+-+-+-+-+-+-+
```

Fig. 4. Modify Route Request Message

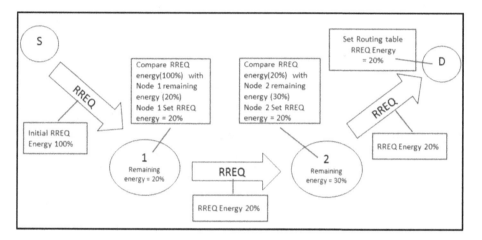

Fig. 5. Set energy in RREQ Message

From Fig. 5, source node (S) sent the information to a Destination node (D). At the beginning source node will broadcast RREQ and set energy parameter in RREQ message to 100%. When node 1 receives RREQ message, it will check the routing table. If the route is not found, node 1 has the energy at 20%. The energy in RREQ has been update to 20% when the energy in Node 1 is less than the energy in RREQ. Then the RREQ is broadcasted to the neighbor nodes. When Node 2 receives RREQ from Node 1, there is no need to update the energy in RREQ due to the remaining energy in node 2 is larger than RREQ. When the destination receives RREQ message from Node 2, it will set energy in routing table and sent RREP back to source node (via node 1,2). If the destination receives multiple RREQ messages in the same time, the algorithm of energy update will be explained in Fig. 6.

The destination receives the first RREQ message from Node 2. After that it receives the second RREQ message from Node 4. In the original AODV, if a destination receives RREQ the same source it will discard the second RREQ message.

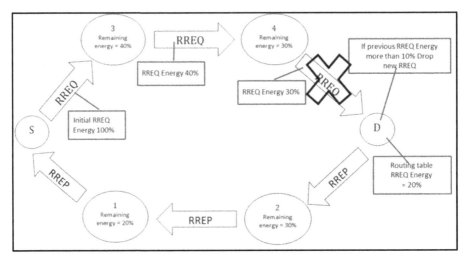

Fig. 6. EW-AODV choose the path when the previous RREQ has RREQ Energy greater than 10%

In EW-AODV, the destination node will check both RREQs and choose the first RREQ if the remaining energy is still more than 10%. However, the destination node will choose another route if the RREQ energy in the first RREQ is less than 10%. In case previous RREQ have RREQ Energy lower than 10% EW-AODV has algorithm to choose path see in Fig. 7.

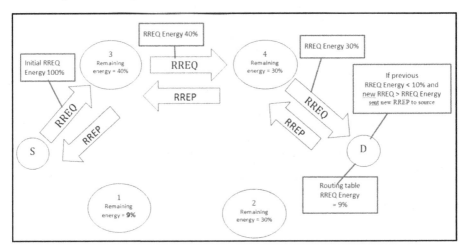

Fig. 7. EW-AODV choose path when previous RREQ have RREQ Energy lower than 10%

From Fig. 7, the destination receives the first RREQ from Node 2 and has RREQ Energy 9% the energy will save in routing table. When the destination receives the second RREQ from Node 4, the RREQ from Node 2 is discarded because the energy in RREQ from Node 2 is less than 10%. In this case, the destination sends a new

RREP to the source node (via node 4, 3). EW-AODV considered the energy parameter to choose the route in contrast with the original AODV. Both EW-AODV and AODV have been simulated using NS-2. The performances in term of throughput, delay, PDR and energy are analyse in the next section.

4 The Simulation

We have compare performance of EW-AODV and AODV base on IEEE 802.15.4 standard with 64 sensor nodes in area 600 m^2,as see in Fig. 8

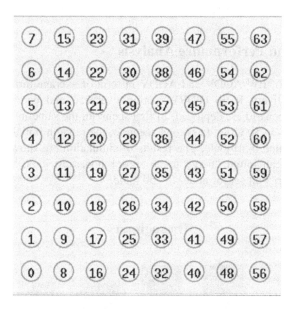

Fig. 8. Set position 64 node

Because of the random position in the simulation, we cannot know the cause of packet drop. The reason of packet drop can be out of transmission rang, collision, run out of energy, no route to send and any reasons. In this research we focus only on the energy cause. The situation has been set for the experiments. Grid topology can make sure that the source has a route to destination. The constant bit rate (CBR) application is set to 64 bytes and data rate at 10 Kbps since a high data rate will have a high probability for collision. Which we don't prefer to occur in the simulation. The simulation time will be 5,000 second and only has one source node in a time. Source will change every 1,000 second in order to force the network changed the route. We set the scenario for comparison between having energy 50 joule (ini) in every node and random energy to nodes in rang of 0-50 joule (rdm). The simulation parameters have set as following:

- Number of Node   : 64 nodes
- MAC   : IEEE802.15.4
- Area   : 600*600 m^2
- Transmission Rang   : 60 m
- Routing Protocol   : EW-AODV, AODV
- Initial Energy   : 50 joule (ini),
                            Random 0-50 joule (rdm)
- Simulation time   : 5000 sec
- Connection   : CBR 1 connection
- Packet Size   : 64 bytes
- Rate   : 10 Kbps

5   Result and Performance Analysis

The performance of EW-AODV and AODV in term of average throughput, average delay, Packet Delivery Ratio (PDR), routing overhead and energy are used for comparison two scenario are deployed. In first scenario, the same initial energy is set to all sensor nodes at 50 joule. In second scenario, the random initial energy is set to all sensor node value 0-50 joule. The results of simulation are show in Fig. 8 – 12. Note that 'ini' mean initial energy to all sensor nodes 50 joule 'rdm' that mean random initial energy all sensor nodes 0 – 50 joule.

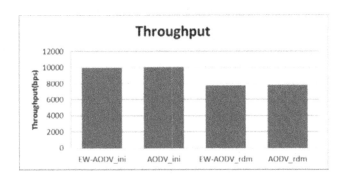

Fig. 9. Average throughput

In the simulation we set the data rate to 10 Kbps, thus the maximum throughput is 10 Kbps in scenario 1. We found that the throughput of both EW-AODV and AODV is not difference. While the random energy initialization gives a lower throughput when compare to the other.

In Fig. 10, the average delay of EW-AODV is slightly greater than AODV because of the discovery modification.

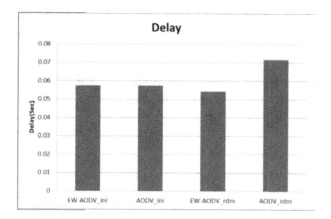

Fig. 10. Average Delay

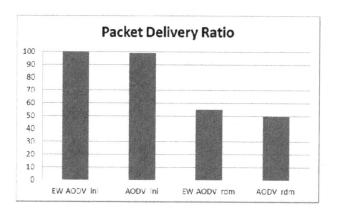

Fig. 11. Packet Delivery Ratio

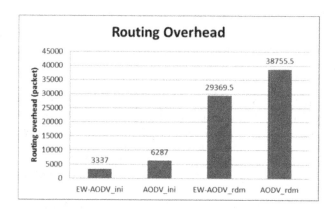

Fig. 12. Routing overhead

In Fig. 11, our EW-AODV produces the best PDR compared with the original AODV. This is because the route is change before running out of the energy. However, the random initial energy of node has less PDR than other.

In Fig. 12, our EW-AODV has a low routing overhead nearly 50% compare to the original AODV in both two scenario. The reason is in EW-AODV, the alternative route is activated before node runs out of the energy. This can avoid the link failure and the network has no need to do a new route discovery.

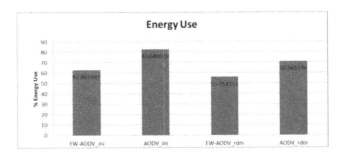

Fig. 13. Energy used

In Fig. 13, we can see our EW-AODV gave the lower energy usage than the random energy. The original in both scenarios. However, our EW-AODV with energy initialization uses the lowest energy at 56%.

6   Conclusion

The simulation result shows that both scenario of EW-AODV and AODV has the similar throughput, PDR and delay. However, our EW-AODV give the good result in term of overhead and energy usage when compare with the original AODV. Therefore our protocol appropriates for wireless sensor networks which energy is its resource constrains.

References

[1] Acs, G., Buttyabv, L.: A taxonomy of routing protocol for wireless sensor networks. Award of the Hungarian Telecommunication Scientific Society, 32–40 (January 2007)
[2] Huang, Z., San-Yang, L., Xiang-Gang, Q.: Overview of Routing in Dynamic Wireless Sensor Networks. International Journal of Digital Content Technology and its Applications 4(4), 199–206 (2010)
[3] Zheng, J., Lee, M.J.: A Comprehensive Performance Study of IEEE 802.15.4. In: Sensor Network Operations, ch. 4, pp. 218–237. IEEE Press (2006)
[4] Perkins, C.E., Belding-Royer, E.M., Das, S.R.: Ad hoc On-Demand Distance Vector (AODV) Routing. RFC 3561 (July 2003)

[5] Felice, D., Relatore, M.: Cross-Layer Optimizations in Multi-HopAd Hoc Networks. Dottorato di Ricerca (March 2008)
[6] Jeongho, S., Tae-Young, B.: A Routing Scheme with Limited Flooding for Wireless Sensor Networks. International Journal of Future Generation Communication and Networking 3(3), 19–32 (2010)

Decision Support in Supply Chain Management for Disaster Relief in Somalia[*]

E. van Wyk, V.S.S. Yadavalli, and H. Carstens

Department of Industrial and Systems Engineering
University of Pretoria
South Africa
{Estelle.vanwyk,Sarma.yadavalli}@up.ac.za,
Hermancarstens@atesouthafrica.com

Abstract. Somalia, a country situated in Eastern Africa has been struggling between rival warlords and an inability to deal with famine. Diseases have resulted to the deaths of up to millions of people. According to a New York Times article on 25 November 2011, Somalia has become a suffering and failed state. The inadequate infrastructure and poorly planned logistics of Somalia may lead to the destruction of the country.

To address these concerns, it is necessary that humanitarian aid is pre-positioned to provide victims with sufficient relief. This chapter addresses some of the issues in supply chain management with the trade-off between stockpile cost and shortage cost by using pre-emptive multi-objective programming. The proposed criteria of the model are described. This is followed by a case study based on Somalia, illustrating the functionality of the model.

1 Introduction

1.1 The Importance of Disaster Management

The severe effects of natural and man-made disasters are made obvious by observing any media source. On 13 November 2011, Agence France-Presse (AFP) stated that United Nations (UN) climate scientists forecast the likelihood of increasing heat waves in Southern Europe [1]. In addition, North Africa will be more susceptible to droughts, and rising seas will cause storm surges in small island states. According to the AFP, peer reviewed scientific journals are claiming that the impact of disasters have a 90% probability of becoming unbearable over time [1]. A summary for policy makers drafted by the AFP claims:

"Global warming will create weather on steroids."

It is feared that in the future, entire communities could be obliterated by a single disaster. The living conditions of communities will degrade as disasters increase in

---

[*] A modified version of this paper was presented at ICMIE conference in Singapore, February 2012

S. Patnaik et al. (Eds.): *New Paradigms in Internet Computing*, AISC 203, pp. 13–22.
DOI: 10.1007/978-3-642-35461-8_2   © Springer-Verlag Berlin Heidelberg 2013

frequency and/or severity, which in turn will cause an increase in permanent migration and present more pressures in areas of relocation, leading to a greater need for disaster management.

Tomasini and Wassenhove [2] define disaster management as the result of a long and structured process of strategic process design, ultimately resulting in successful execution. Disaster management can be divided into four phases: mitigation, preparedness, response and recovery. These phases are known collectively as the disaster operations life cycle. Mitigation is the application of measures that either prevent the onset of a disaster or reduce the impact should a disaster occur. Preparedness relates to the community's ability to respond when a disaster occurs; response refers to the employment of resources and emergency procedures as guided by plans to preserve life, property, and the governing structure of the community. Finally, recovery involves actions taken to stabilize the community subsequent to the immediate impact of a disaster [2]. The disaster cycle is illustrated in Figure 1[1].

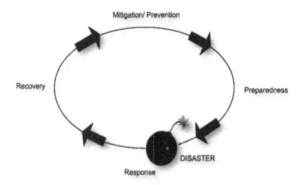

Fig. 1. The disaster cycle

Tomasini and Wassenhove [2] emphasise that the first 72 hours after a disaster has occurred are crucial in order to save the maximum amount of human lives. Saving lives, however, relies on the correct quantity and types of aid supplies, which would be a fairly effortless resolution, if all disaster effects could be predicted. Arminas [4] suitably describes this predicament as follows:

> "...purchasing and logistics for major disaster relief is like having a client from hell: You never know beforehand what they want, when they want it, how much they want and even where they want it sent."

As a consequence of this complexity, it is vital that relief supplies be pre-positioned to improve emergency response times. This forms part of the preparedness phase in the disaster operations life cycle. Demand for aid supplies will vary in type and quantity depending on the specific disaster and the level of destruction it causes. These supplies

---

[1] Adapted from Ciottone [3].

must meet the immediate needs of those affected and will include items such as food, medicine, tents, sanitation equipment, tools and related necessities [5].

Considering the problem of preparing for a disaster, Somalia is suffering severely due to poverty, famine and the strongholds of the terrorist group, al-Shabaab. Somalia is in urgent need of humanitarian aid and is therefore considered as the case study for this work. The existing disasters for Somalia are identified as well as the associated inventory which would be necessary for survival.

1.2   Somalia as a Case Study

Since the spring in 2011, Somalia experienced a drought which is considered to be the worst in 60 years [6]. The country has suffered from crop failure, an extreme rise in food prices, as well as the grip of al-Shabaab on central and South Somalia. These factors have forced the United Nations (UN) to declare famine in six areas of Somalia. These areas have currently been reduced to three areas due to the assistance of humanitarian organizations over the past few years. The increase in assistance, however, has been insufficient as there are still 3.7 million victims in need of emergency assistance and 250,000 in danger of dying. The famine has compelled thousands of victims to move to overfull refugee camps in Ethiopia, Kenya, and Djibouti, while other victims have fled to Internally Displaced Persons camps in Mogadishu [6]. The United States, UN, and international humanitarian organizations have been trying to address the immediate needs of the victims in Somalia [6].

The unfortunate circumstances leaves Somalia in need of a more permanent solution, i.e., more research should be done so that adequate solutions are obtained and facilities should be pre-positioned in disaster areas with a sufficient supply of relief aid. This study is based on the formulation of mathematical modeling as a means to determine the quantities of aid supplies at a pre-positioning facility within Somalia. Preemptive multi-objective programming is employed to ensure that areas will be supplied with immediate disaster relief.

The remainder of this chapter is structured as follows: A literature review addresses the problem variants associated with disaster relief and existing models that have been developed to accommodate these areas. The research methodology presents the formulated model. The model results and the findings are then discussed, followed by some conclusions and future research developments.

2   Literature Review

2.1   Problem Variants

Considerable literature has addressed the management of disaster relief organizations. Much of this deals with the social and organizational implications of responding to disasters in many parts of the world, including countries with poor infrastructure and/or may be involved in hostilities. Blecken et al. [7] state that even though research contributions to supply chain management in the context of humanitarian operations have increased, a gap remains when considering pre-positioning in countries such as Somalia. Being prepared for a disaster requires the knowledge of knowing when or where an event will take place, how many people will be affected and what supplies will be required. Despite the progress that disaster planning, mitigation and new management systems have made, the need for relief, specifically in underdeveloped countries, still remains [5]. Improving disaster relief planning and management is a continuous process.

Due to the unpredictable nature of a disaster, disaster management is a process that cannot be comprehensively controlled. Altay and Green [8] explain that even though it is known that response to disasters requires effective planning, it is crucial to leave room for improvisation in order to deal with the unusual challenges that manifest. Hills [9] approvingly states that the phrase disaster management implies a degree of control, which rarely exists in disaster cases. It is for this reason that Standard Management Methods used in industry may not always apply directly to disaster situations [9].

Rawls and Turnquist [10] raise an added concern, namely that the capacities of resource providers are the key components in managing response efforts subsequent to disaster events, but that only a small amount of research has been conducted on the planning and distribution of aid supplies kept in inventory at prepositioned facilities. In addition, Duran et al. [11] maintain that an important element to take into account when considering stock pre-positioning is that facilities should always have sufficient inventory to satisfy demand. It should also be considered that stored aid supplies may be destroyed during a disaster event [11]. The pre-positioned stock should thus meet the needs of a disrupted region by taking the effect of the disaster into consideration [12].

Any shortcomings may result in serious consequences for victims of disasters and could mean the difference between life and death [2]. The public thus expects "perfect orders" and that humanitarian supply chains need to be more adaptable and agile towards the changing needs of disaster victims [2]. This need demands effective methods to improve disaster preparedness.

2.2   Existing Solutions

The majority of favourable solutions to disaster management problems are supported by mathematical methods such as operations research [13]. This approach is an appropriate tool for planning the preparedness, response and recovery phases of disaster management, due to its ability to handle uncertainty by means of probabilistic scenarios which represent disasters and their outcomes [14, 15, 16, 17].

Very few journal articles address logistical problems that are related to humanitarian relief. Rawls and Turnquist [10] present a two-stage Stochastic Mixed Integer Program(SMIP) that provides an emergency response pre-positioning strategy for disaster threats. The algorithm is formulated as a heuristic algorithm. The model considers uncertainty in demand for stocked supplies but also includes the uncertainty regarding transportation network availability after an event. A stochastic inventory control model is developed by Beamon and Kotleba [18] in the form of $(Q_1;Q_2; r_1; r_2)$. The model approach is to use optimal order quantities and re-order points to determine inventory for a pre-positioned warehouse responding to a complex humanitarian emergency, including the exceedingly variable demand of the warehouse supply items [18]. The model allows for two types of order lot sizes: Q_1 for a regular order and Q_2 for an urgent order. Q_1 is ordered when the inventory reaches level r_1 and Q_2 is ordered when the inventory level reaches r_2, where $r_1 > r_2$.

A Markovian process is also used to solve the demand distribution of inventory. This idea is initiated by Karlin and Fabens [19], claiming that if each demand state is defined by different numbers, a base stock type inventory policy can be obtained. Taskin and Lodree [20] use stochastic programming to determine an optimal order policy so that the demand in each pre-hurricane season period is met, and reserve supplies are stored for the ensuing hurricane season, in a cost-effective way.

Bryson et al. [12] use optimal and heuristic approaches to solve a number of hypothetical problems. Mixed integer programming is applied to establish the disaster recovery capability of an organisation. The aim of the model is to determine the resources that should be used in order to maximise the total expected value of the recovery capability. The use of mathematical modelling provides an appropriate decision support tool for the successful development of a Disaster Recovery Plan (DRP). This model provides a generic approach which considers different types of resources required to satisfy demand induced by any relevant disaster.

Various models have been developed and applied to the SADC countries [13,20]. VanWyk et al. [13] apply a stochastic inventory model to the SADC countries to obtain the quantities of aid supplies to keep at an acceptable minimum cost. Van Wyk et al. [22] developed a mixed integer decision model which selects sub-plans to supply a country with immediate disaster relief. In addition, a Euclidean Distance Algorithm was formulated to determine the most similar case when compared to a target case disaster [22].

The research done for disaster management problems and specifically the SADC applications, provide useful methods, but only limited research considers case studies applicable to developing countries. This research is aimed at focusing on the capacity required to ensure that a sufficient area within Somalia is covered by a pre-positioning facility. It is therefore a good starting point to develop an innovative model to determine the capacity required to ensure that a sufficient area within Somalia is covered by the pre-positioned facility.

2.3   Preemptive Multiobjective Programming

Rardin [23] explains that although practical problems almost always involve more than one measure of solution merit, many can be modelled quite satisfactorily with a single cost or profit objective. Other criteria are either presented as constraints or weighted in a composite objective function to produce a model efficient enough for productive analysis. Many applications such as those in disaster management must be treated as multiobjective. When goals cannot be reduced to a similar scale of cost or benefit, trade-offs have to be addressed. To obtain useful results from such a problem, the multiobjective model must be reduced to a sequence of single objective optimizations [23]. This leads to preemptive multiobjective optimization by considering objectives one at a time. The most important objective is optimized subject to a requirement that the first has achieved its optimal value; and so on [23].

The preemptive approach to multiobjective optimization is that it results in solutions that cannot be improved in one objective without degrading another. If each stage of the preemptive optimization yields a single-objective optimum, the final solution is an efficient point of the full multi-objective model. The preemptive process uses one objective function at a time to improve one without worsening others. At the completion of this process, no further improvement is possible. As usual, infeasible and unbounded cases can produce complications, but the typical outcome is an efficient point [23].

3   Research Methodology

3.1   Preemptive Optimization

Humanitarian relief organizations aim to provide relief for as many disaster victims as possible, subject to limited funding. It is therefore useful to consider a model that helps the decision-maker with inventory decisions at the lowest possible cost. The notation of the preemptive model [6] for Somalia is addressed below:

x_{ik} $\triangleq$ $\begin{cases} 1 \text{ if aid supply i is required for disaster k} \\ 0 \text{ otherwise} \end{cases}$

q_k $\triangleq$ The probability that disaster k will occur

n_k $\triangleq$ The number of people affected by disaster k

c_i $\triangleq$ The unit cost of aid supply i

h_i $\triangleq$ The holding cost of supply i

s_i $\triangleq$ The number of people affected if supply i is not available

u_i $\triangleq$ The number of people affected if supply i is not available

v_i $\triangleq$ The number of aid supply i in excess

Q_i $\triangleq$ The number of aid supply i required

The objective functions have been formulated as follows:

$$min \ Z_1 = \sum_{i=1}^{I} \ Q_i c_i + h_i v_i \tag{1}$$

$$min \ Z_2 = \sum_{i=1}^{I} \ s_i u_i \tag{2}$$

s.t.

$$Q_i - v_i + u_i = \sum_{k=1}^{K} \frac{x_{ki} n_k q_k}{s_i}, \ i \in I \tag{3}$$

$$Q_i, v_i, u_i \geq 0 \qquad (4)$$

Objective function (1) minimizes the overall cost of holding excess aid supplies. Objection function (2), minimizes the shortage cost (number of lives affected) of not having an aid supply. Constraint (3) guarantees that the number of aid supplies required for a specific disaster corresponds with the expected demand of a scenario, while taking excess inventory and shortages into consideration. Constraint (4) ensures that decision variables Q_i, v_i and u_i remain greater or equal to 0. It is assumed that no excess inventory is present during the first usage of the model.

3.2 Data Gathering

Predicting a disaster is challenging, and in most cases impossible. However, a probability can be determined to pre-determine the likelihood of such an event. The approach used to determine these estimates was to observe the number of times the identical disasters have occurred in Somalia in the past 30 years. We use data from the Emergency Disaster Database (EM-DAT) as provided by the Centre for Research on the Epidemiology of Disasters (CRED) [24]. In this database, an event qualifies as a disaster if at least one of the following criteria are fulfilled: 10 or more people are reported killed; 100 or more people are reported affected, injured and/or homeless; there has been a declaration of a state of emergency; or there has been a call for international assistance. We measure the severity of a disaster in Somalia by the number of people affected.

The repetition of occurrences of each disaster is then divided by the overall total of Somalia disasters, giving the result of q_k of each disaster. The parameter n_k, represents the estimated number of victims likely to be affected by a disaster in its worst magnitude. Therefore, if a drought occurs, it is most likely that the entire population (100%) may be affected. These values are multiplied by the total population of an area to give an indication of the total victims affected.

Holding cost comprises the cost of carrying one unit of inventory for one time period, and usually indicates storage and insurance cost, taxes on inventory, labor cost, and cost of spoilage, theft, or obsolescence [25]. However, the inventory carrying cost will vary according to each individual warehouse, but for testing purposes it is assumed that inventory carrying cost equals 25% of product value per annum [13]. The shortage cost represents the amount of people who will be affected if an aid supply is not available during and after the disaster event.

The preemptive optimization model performs multi-objective optimization by first optimizing objective function (1), i.e., the cost of holding an aid supply. Objective function (2), i.e., the cost of the total shortages, is optimized subject to the requirement that 1 has achieved its optimal value [23].

4 Results and Findings

The model was solved to construct four efficient frontier curves, each representing a category. The efficient frontier indicates the efficient points when considering the holding and shortage cost for each category. The efficient points and the efficient frontier assist to characterize "best" feasible solutions in multiobjective models. Category A illustrates the efficient frontier for 0 – 1 million people affected, category B between 1 million – 2 million people, category C between 2 million – 3 million people and category D between 3 million and 4 million people affected. The four categories are given in Figure 2, illustrating that with each category the number of aid supplies will increase, increasing the overall costs of a pre-positioning facility. The categories can be used as a decision tool to determine the quantities of supplies to be kept within an acceptable budget.

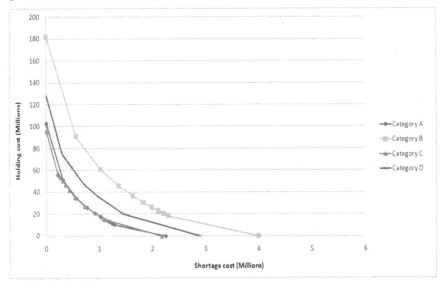

Fig. 2. Efficient frontier for each category

5   Conclusion and Future Work

The aim of this paper is to show how mathematical modeling can provide strategic decision support for selecting the required amount and types of aid supplies, and the most appropriate quantities of pre-positioning facilities within Somalia.

The models address the concerns of feasibility, consistency and completeness. From a decision-maker's point of view, the model can serve as a convenient guideline to assist with planning of types and quantities of aid supplies that should be kept for sufficient preparedness. The results of the model indicate that workable solutions have been identified, which have unveiled the possibility to increase the use of operational research methods to enhance disaster relief decision making.

Future research developments can complement this work by gathering data from other countries and applying the models to the pre-selected regions. It is also necessary to optimal man-power required during and after a disaster has occurred and finally, risk factors need to be incorporated as part of the formulation.

In conclusion, this research challenges the fatal effects of disasters by providing instruments to overcome some of the difficulties of disaster management. Swami Vivekananda cautions [26]:

> "If you think about disaster, you will get it. Brood about death and you hasten your demise. Think positively and masterfully, with confidence and faith, and life becomes more secure, more fraught with action, richer in achievement and experience."

Disasters cause a great deal of suffering, but through careful planning, evaluation and the implementation of ongoing research, it is possible to achieve the aim of improving disaster preparedness through keeping sufficient amount of aid supplies in pre-positioned facilities in Somalia, at reasonable and affordable cost structures.

Acknowledgments. The authors thank the National Research Foundation and the University of Pretoria for the financial support.

References

[1]   News 24. More weather extremes expected (2011), http://m.news.24.com/ (accessed November 13, 2011)

[2]   Tomasini, R., Wassenhove, L.: Humanitarian Logistics. Palgrave Macmillan, U.K. (2009)

[3]   Ciottone, G.R.: Disaster Medicine, 3rd edn. Elsevier Mosby, U.S.A. (2006)

[4]   Arminas, D.: Supply lessons of tsunami aid. Supply Management. The Purchasing and Supply Website 10(2), 14 (2005)

[5]   Whybark, D.: Issues in managing disaster relief inventories. International Journal of Production Economics 108, 228–235 (2007)

[6]   Bureau of African Affairs. U.S. Department of State, Somalia (2012), http://www.state.gov/r/pa/ei/bgn/2863.htm (accessed February 10, 2012)

[7]   Blecken, A., Danne, C., Dangelmaier, W., Rottkemper, B., Hellinggrath, B.: Optimal Stock Relocation under Uncertainty in Post-disaster Humanitarian Operations. In: Proceedings of the 43rd Hawaii International Conference on System Sciences, pp. 1–10. IEEE Computer Society Press (2010)

[8]   Altay, N., Green, W.G.: OR/MS research in disaster operations management. European Journal of Operations Research 175, 475–493 (2006)

[9]   Hills, A.: Seduced by recovery: The consequences of misunderstanding disaster. Journal of Contingencies and Crises Management 6(3), 162–170 (1998)

[10] Rawls, C.G., Turnquist, M.A.: Pre-positioning of emergency supplies for disaster response. Transportation Research, Part B 44(4), 521–534 (2009)

[11] Duran, S., Gutierrez, M.A., Keskinocak, P.: Pre-positioning of Emergency Items Worldwide for CARE international. INFORMS (2009) doi: 10.1287

[12] Bryson, K.M.N., Millar, H., Joseph, A., Mobolurin, A.: Using formal MS/OR modeling to support disaster recovery planning. European Journal of Operations Research 141, 679–688 (2002)

[13] Van Wyk, E., Bean, W.L., Yadavalli, V.S.S.: Modelling of uncertainty in minimizing the cost of inventory for disaster relief. South African Journal of Industrial Engineering 22(1), 1–11 (2011a)

[14] Mete, H.O., Zabinsky, Z.B.: Stochastic optimization of medical supply location and distribution in disaster management. International Journal of Production Economics 126, 76–84 (2009)

[15] Snyder, L.V.: Facility location under uncertainty: a review. IIE Transactions 38(7), 547–564 (2006)

[16] Özdamar, L., Ekinci, E., Küçükyazici: Emergency logistics planning in natural disasters. Annals of Operations Research 129(1-4), 217–245 (2004)

[17] Beraldi, P., Bruni, M.E.: A probabilistic model applied to emergency service vehicle location. European Journal of Operations Research 196(1), 323–331 (2009)

[18] Beamon, B.M., Kotleba, S.: Inventory modeling for complex emergencies in humanitarian relief operations. International Journal of Logistics: Research and Applications 9(1), 1–18 (2006)

[19] Karlin, S., Fabens, A.: The (s,S) inventory model under Markovian demand process. Mathematical Methods in the Social Sciences, 159–175 (1960)

[20] Taskin, S., Lodree, E.: Inventory decisions for emergency supplies based on hurricane count predictions. International Journal of Production Economics, 1–10 (2009)

[21] Van Wyk, E., Yadavalli, V.S.S., Bean, W.L.: Strategic inventory management for disaster relief. Management Dynamics 20(1), 32–42 (2011b)

[22] Van Wyk, E., Yadavalli, V.S.S.: Application of an Euclidean distance algorithm for strategic inventory management for disaster relief in the SADC. In: Proceedings of the 41st International Conference on Computers & Industrial Engineering, October 23-27, Los Angelas, pp. 894–899 (2011c)

[23] Rardin, R.: Optimization in Operations Research. Prentice Hall, Upper Saddle River (1998)

[24] CRED, Country Profiles (2009), http://www.emdat.be/disaster-profiles (accessed January 10, 2012)

[25] Winston, W.: Introduction to probability models, 4th edn. Curt Hinrichs (2004)

[26] Swami Vivekananda, Think Exist (2010), http://www.thinkexist.com/ (accessed March 22, 2010)

Challenges in Cloud Environment

Himanshu Shah, Paresh Wankhede, and Anup Borkar

Accenture Services Private Limited,
Mumbai, India
{himanshu.shah,paresh.wankhede,anup.borkar}@accenture.com

Abstract. Organizations have been skeptical about moving from traditional data-centre onto cloud. Along with well known concerns like security, loss of control over infrastructure, data theft, lack of standards, etc; cloud does pose issues like portability, reliability, maintainability, ease-of-use, and etcetera.

This whitepaper talks about these concerns around system quality attributes using Amazon Web Services (AWS) and Azure cloud as reference. The whitepaper encompasses the recent challenges faced and probable solutions for the same. It also covers one specific issue related to RHEL (Red Hat Enterprise Linux) [Ref 3] migration on AWS in detail.

This whitepaper also discusses and recommends cloud vendor(s) and certain management tools based on the parameters related to system quality attributes such as portability, reliability, maintainability, etc.

1 System Quality Attributes

1.1 Portability

It is the ease with which an environment and an application can be ported on and off the cloud or to other public cloud.

a) *Description:* There are no common standards for cloud vendors to adhere to. Current development efforts do not suffice the purpose. And they do not force the cloud service providers to prioritize and focus on the issue of interoperability. Some examples of it are:

1. Microsoft's Azure only supports Windows OS compared to other vendors like AWS which support various flavors of Unix/Linux.
2. A java application hosted in Google Apps Engine (GAE) is bound to the DataStore which isn't exactly an RDBMS.
3. AWS PaaS services like SQS [Ref 4], RDS [Ref 5] creates a vendor lock-in.

All the cloud vendors have the liberty to implement their services the way they deem beneficial. AWS provides a bunch of loose coupled services which can be used either in conjunction or independently. AWS, as IaaS, does not force the user to change the architecture of an application to make it compatible to host on AWS cloud. So at any

S. Patnaik et al. (Eds.): *New Paradigms in Internet Computing,* AISC 203, pp. 23–30.
DOI: 10.1007/978-3-642-35461-8_3 © Springer-Verlag Berlin Heidelberg 2013

given point in time user has a flexibility to move out of AWS cloud to any other cloud or traditional data-centre. Though there are PaaS services provided by AWS, if used, would require change in application architecture and that in turn would mean vendor lock-in.

b) *Recommendation*: There are no open standards. Cloud users have to keep this constraint in mind while designing the application, so that they don't get locked-in with the vendor. AWS provides support for application portability as long as PaaS services are not used.

1.2   Recoverability

It's the ease with which the infrastructure and application could be recovered in case of a disaster and/or threat to Business continuity.

a) *Description:* Services provided by cloud vendors are unique to their implementation. The compute node management, back-up procedures, firewall implementation, is different for all the vendors. Considering these facts application recoverability could be a time consuming.

It is quintessential in today's fast growing business environment to automate IT provisioning. Managing and configuring the IT infrastructure is one of the most time consuming and error prone task. The desire to implement something that would make the paradigm shift has given birth to a concept which treats Infrastructure management as Code. This transition, from traditional server maintenance to automation, would make the building and maintaining a modern infrastructure environment look more like maintaining a software project. AWS provides a service called CloudFormation [Ref 6] using which a user can rebuild the whole of infrastructure right from the scratch without much of human intervention.

Treating the Infrastructure as Code has been widely acknowledged. There are certain service providers/technologies which offer Infrastructure management. They are:

1. Chef [Ref 7]
2. Puppet [Ref 8]
3. RightScale [Ref 9]
4. Eucalyptus [Ref 10]

b) *Recommendation:* It is recommended that Chef or Puppet should be used be it any infrastructure environment. These tools come handy for the purpose of recoverability and maintainability. AWS and Azure along with cloud management tool provides a unique way to manage cloud servers. The pros and cons of each of these tools have been discussed in details in coming sections.

1.3   Maintainability

It is the ease with which the application environment could be maintained.

a) *Description:* One of the biggest motivation factors for organizations to move onto cloud from traditional data-centre is maintainability. Users need not to worry about the maintenance of infrastructure when on cloud. Though, users have to keep a tab on application's health and status. This can be done with the help of cloud management tools available. Chef and Puppet are specially built considering this requirement. These tools help maintain the required status of an application. All the leading cloud service providers support cloud management through APIs, CLIs and web consoles. For example, AWS provides Java APIs while Azure supports .Net APIs.

To discuss the maintainability an experience of RHEL migration on AWS cloud has been discussed in detail in subsequent section.

b) *Recommendation:* Though there are bunch of APIs and CLIs available with each of the cloud vendor; it is easy to manage and maintain the infrastructure on cloud with the help of cloud management tools.

1.4   Transparency and Control

It is the extent to which the cloud vendor allows its user to control the infrastructure.

a) *Description:* Transparency in cloud's perspective is the capability to have a view at day-to-day activity of cloud infrastructure or point-in-time status and health of the servers. Users would like to have full control over the servers which are running under their account and hosting their applications. AWS implementing IaaS gives the user right amount of control over the servers to suffice the purpose. Azure on the other hand being PaaS does not provide that level of transparency and control at the hardware level.

b) *Recommendation:* If transparency and control over infrastructure is too major a deciding factor, data-centre is a way to go. AWS provides some control over the infrastructure which should suffice in most of the cases.

1.5   Security

It is the extent to which the cloud vendor supports security and reliability for cloud environment.

a) *Description:* One of the deciding factors for any organization while choosing cloud environment is its security and reliability. There are numerous regulations appertain to the storage and use of data, including Payment Card Industry Data Security Standard (PCI DSS), ISO, the Health Insurance Portability and Accountability Act (HIPAA), the Sarbanes-Oxley Act, to name a few.

Beside this there are privacy issues that arise from virtualization. AWS runs the server on virtualized environment and this brings in the issue that could arise from multi-tenancy. Organizations are skeptical about sharing the same hardware with multiple users. Though there are few options such as virtual private cloud and virtual private network that helps resolve the data security and network security to some level.

b) *Recommendation:* There are provisions that could help resolve the security issues to some extent, but if the nature of data in question is very sensitive it is better to keep it on on-premise servers.

1.6   OS Support

It is the number and verity of operating systems that are supported.

a) *Description:* AWS supports almost all the types of leading OS [REF 1] ranging from enterprise level Unix systems to Microsoft server OS.
b) *Recommendation:* AWS has various flavors of operating systems to choose from. That makes a viable contender for application hosting.

2   RHEL Migration on AWS Cloud – An Experience

AWS along with Red Hat offers enterprise level operating system on cloud. AWS-Red Hat collaboration allows users to rent Red Hat Enterprise Linux instead of purchasing a license. With this combined effort users were able to:

a) Purchase RHEL by paying hourly fees.
b) Use RHEL OS which was provided on Amazon EC2 [Ref 1] cloud.
c) Get the supported from Red Hat.

To sum it all up; the images were provided by AWS EC2 cloud, supported by Red Hat and managed by end user (Ref Figure 1 below).

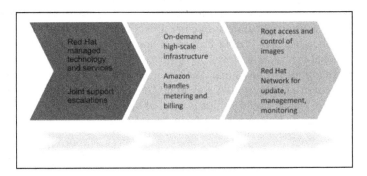

Fig. 1. The Fuzzy frequent pattern tree

The EC2 images provided by Red Hat were available in RHEL version 5.x. There were following ways to subscribe to that service:

a) On-demand pay-as-you-go
b) RHEL premium subscription

This was the beta offering which has now been discontinued to introduce hourly on demand offering. Subsequent section talks about the underlying architecture of AWS on top of which RHEL is offered.

2.1   Amazon as-is Architecture

All the services provided by AWS are based on virtualization and its computing unit EC2 is no exception. Following subsections discuss in detail the kernel level architecture of AWS with respect to RHEL.

a) *Amazon Amazon Kernel Image (AKI) and Amazon RAM Image (ARI) Concept*

Amazon gives the flexibility to use the kernels of user's own choice, other than the default EC2 kernels (AKI). To complement these kernel images AWS also provides ARI. The point to be noted here is that the architecture of the selected AMI, AKI and ARI should match and be compatible [Ref 2]. Even if they match, there is no guarantee that the combination would work and server would instantiate. It could so happen that the combination in place is not meant or designed to work in conjunction.

b) *Para-Virtual GRUB Loader (PVGRUB)*

AWS EC2 has a provision to load a Para-virtual Linux kernel within an Amazon Machine Image (AMI). The option is provided to create images that contain a kernel and initrd (initial RAM disk), and behave in a manner that is closer to traditional virtual or physical Linux installations. This feature allows seamless upgrade of kernel on Amazon EBS-backed instances. It enables the user to load any kernels of their choice. This PVGRUB acts as a mini-OS that runs at the boot and selects the kernel to boot from by reading /boot/grub/menu.lst from an AMI. It loads the specified kernel and eventually shuts itself down to free the resources.

2.2   Migration from DevPay to Hourly on Demand

Initially, RHEL (beta image) was made available on EC2 via DevPay subscription model. Users could simply subscribe to the offering, electronically, and have access to the RHEL AMIs. The subscription came with monthly fees on top of pay-per-use (hourly fee) model of AWS.

Red Hat has recently made an announcement to perform a major AMI update. This update will remove all the existing DevPay AMIs and replace them with the new set up AMIs. This move has been introduced to change the subscription model from DevPay to hourly on demand. This new subscription model does not include any monthly fees.

There is no migration tool or policy made available by RHEL or AWS. Considering the fact that DevPay AMIs would not work post the retirement, all the data from the configured AMIs had to be moved onto the new offering images called hourly on demand. There are two ways to get this done:

a)   Doing the whole configuration right from the scratch.
b)   Utilizing available cloud management tools.

Subsequent section details out the theory of utilizing cloud management tools, which is the recommended way to manage the infrastructure. These management tools are based on a concept arisen from treating the Infrastructure as Code.

2.3   Infrastructure as Code

It is quintessential in today's fast growing business environment to automate IT provisioning. Managing and configuring the IT infrastructure is one of the most time consuming and error prone task. The desire to implement something that would make the paradigm shift has given birth to a concept which treats Infrastructure management as Code. This transition, from traditional server maintenance to automation, would make the building and maintaining a modern infrastructure environment look more like maintaining a software project.

Using cloud management tools gives an edge over traditional methods in a sense that they are agile, modular and customizable. This need of managing the infrastructure and automating the provisioning process has made the users to look for a solution which could be perceived as a software module. Cloud management tools such as Chef and Puppet are open source scripting based management tools. These scripts are idempotent, cross platform and modular which makes them a reliable to use.

Various cloud management tools are compared and discussed in following sections [Ref Table 1]. Recommendations based on predefined parameters are also provided.

3   Recommendations

3.1   Manageability

- *Script the image configuration process*
  Scripting the instance configuration process gives an edge over manual configuration process in an ever changing environment (refer the RHEL case discussed above). It's easier to modify a piece of code than repeating the manual process right from the scratch.

3.2   Configurability

- *Assigning a role to an instance on the fly*
  It's always good to use the base image for instance creation and configure that instance, by assigning it a role, on the fly. This reduces the burden of maintaining the bunch of images. Any changes that are to be made would have to be incorporated in the base image.

3.3   Maintainability

- *Usage of cloud management tools*
  Puppet and Chef are two such cloud management tools which offers infrastructure automation. They provide a framework to automate system admin tasks. There are certain benefits of using scripting tools to set up the infrastructure against traditional, manual method. The advantages are as follows:
  a) Easier to maintain.
  b) No scope for errors inflicted by human intervention.
  c) Same scripts could be used in traditional datacenter.
  d) No vendors lock-in as far as cloud vendors are concerned.
  e) Scripts are idempotent, cross platform and modular.

There are some negatives associated with this. They are as follows:

a) Big learning curve.
b) Requires deep knowledge and understanding of how system would behave if code were to change even a bit.

AWS provides a service called CloudFormation which actually works as cloud management tool to recreate whole of the infrastructure with a single click. This is based on a concept of Infrastructure as Code. Azure does not provide any such functionality right out of the box

- *Third party cloud management tool*
  Using a third party management tool, like RightScale, gives a flexibility to move the application from one cloud service provider to another with ease. But there is a possibility of getting tied up to RightScale services. RightScale has its own terminology and features like ServerTemplate, RightScripts which are very handy and useful, but are RightScale patented.

3.4   Portability

- *Avoid vendor lock-in*
  An application to be ported on cloud should be designed such that it becomes cloud agnostic. Services that could make the user get locked-in with vendor should be avoided to make the application portable in future. Such applications could make use of cloud management to its full extend. Services provided by Azure are more tightly coupled as compared to services provided by AWS. There are high chances of vendor lock-in with both the service providers if specific services like queuing service, Content Delivery Network (CDN), etcetera, are used.
  Following table (Table I) lists out the recommendation for different cloud management tools based on system quality attributes:

Table 1. Comparison And Recommendation On Cloud Management Tools

Capabilities	RightScale	CloudFormation	Chef/Puppet	Eucalyptus	Recommendation
Cloud vendor Portability	Supports various clouds, but not all of them.	Specific to AWS cloud	Scripts works on any cloud and Data-centre.	Supports AWS and on-premise cloud.	Chef/Puppet.
Recoverability	It is helpful	Built for recoverability	Supports recoverability	Supports recoverability	CloudFormation.
Usability	Small learning curve considering the fact that it has a web console	Small learning curve. Based on JSON	Big learning curve	It has a learning curve.	CloudFormation.

Table 1. (*continued*)

	Easy to maintain	Easy to maintain	Somewhat difficult	Easy to maintain	RightScale.
Maintainability	Easy to maintain	Easy to maintain	Somewhat difficult	Easy to maintain	RightScale.
Lock-in	User gets tied up to RightScale	Only usable with AWS	No lock-in	User gets tied up to Eucalyptus	Chef/Puppet
User group	Big user group. Components are sharable.	No formal platform to share components	Big user group. Components are sharable.	Small user community.	RightScale or Chef/Puppet
Safety	Have to share cloud credentials.	No need to share credentials	No need to share credentials	Have to share cloud credentials.	Chef/Puppet

4   Conclusion

System quality attributes are the key parameter on which maturity of a cloud service can be evaluated. Cloud computing still being in a nascent state strives to provide quality service over this attributes. Evaluation and recommendation in this whitepaper is based on current state and maturity of various clouds and cloud management tools. This is subjected to change down the line, considering the fact that cloud technologies are in evolving state.

Treating Infrastructure as Code for cloud environment would reduce an overhead to configure and manage the infrastructure manually. Same scripts, with little or no modifications, could be used to setup an infrastructure in a datacenter. These scripts/tools are not vendor specific and hence provide a great flexibility and agility. Scripts are idempotent, cross platform and modular which makes them a reliable to use.

RHEL migration case discussed in this whitepaper is an apt example on maintainability issue that one could face on cloud and it also highlights the importance of treating the Infrastructure as Code.

References and Further Reading

[1] http://aws.amazon.com/ec2/
[2] http://aws.amazon.com/articles/1345
[3] http://aws.amazon.com/rhel/
[4] http://aws.amazon.com/sqs/
[5] http://aws.amazon.com/rds/
[6] http://aws.amazon.com/cloudformation/
[7] http://www.opscode.com/chef/
[8] http://puppetlabs.com/
[9] http://www.rightscale.com/
[10] http://www.eucalyptus.com/

Challenges on Amazon Cloud in Load Balancing, Performance Testing and Upgrading

Himanshu Shah, Paresh Wankhede, and Anup Borkar

Accenture Services Private Limited,
Mumbai, India
{himanshu.shah,paresh.wankhede,anup.borkar}@accenture.com

Abstract. Web application hosting in a data centre is clouded with quite a few issues ranging from hardware provisioning, software installation and maintaining the servers. Traditional data-centre techniques need production grade hardware to test application's behavior/performance under expected peak load. This could be costly and procuring hardware could be very time consuming causing delays in software delivery. Cloud (Infrastructure-as-a-Service) can be an answer to this. Cloud Computing provides production grade server instances at very cheap rates.

This whitepaper is divided into two sub parts: first part details out the typical web application setup on Amazon Web Services cloud (AWS) [Ref 2], challenges faced during the setup and resolution for the same, while the second part talks about the observations made during load testing using Apache JMeter performance testing tool on AWS cloud. Three different application setup topologies (single tier, two tier and three tier) are tested and findings and learning from it are discussed here.

This whitepaper only highlights the pitfalls encountered and possible resolutions for each and is not a comment on performance of Amazon cloud. The whitepaper endeavors to find out the best architecture which would give maximum return on investment.

1 Hosting a Web Application on Cloud

Organizations are lured into moving onto cloud from traditional data centre to reap the benefits of its agility, elasticity, cost-efficiency and security.

1.1 Issues with Traditional Infrastructure

There are some issues with traditional infrastructure which increases application's 'time-to-market' parameter. Those issues could be:

a) *Hardware provisioning*
b) *Installation of software*
c) *Prediction of expected usage/load*
d) *Scale up and Scale down*
e) *Hardware maintenance*

S. Patnaik et al. (Eds.): *New Paradigms in Internet Computing,* AISC 203, pp. 31–40.
DOI: 10.1007/978-3-642-35461-8_4   © Springer-Verlag Berlin Heidelberg 2013

The best way to deal with these could be to move the application on Cloud. The subsequent sections talk about deploying a typical web application on AWS cloud with various topologies. Along with it are the recommendations based on performance parameter.

1.2   Typical Web Application Topology

A typical web application could take one of the following forms:

a) *Single tier (all in one)*
b) *Multi tier*

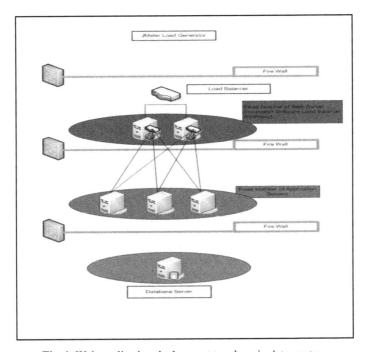

Fig. 1. Web application deployment topology in data-centre

In a typical data-centre, hardware sizing is predefined according to the expected peak request load. This approach generally leads to over-sizing or under-sizing of the hardware. Apart from this, server hardware and software maintenance is an overhead. AWS cuts through these traditional challenges to provide elasticity and agility. AWS reduces the capital expenditure required for initial hardware setup (AWS provides pay-as-you-go model with no upfront investment).

1.3   Web Application on AWS

Organization has been skeptical about application migration on cloud. Reasons could be one or more of the following:

a) *Data security on public cloud*
b) *Performance issue of an application*
c) *Losing control over infrastructure*
d) *Vendor lock-in*

There are various documents and whitepapers that talk about the data security on cloud. Performance testing of a web application on cloud is something that has not been touched upon. We have addressed this topic by deploying a simple web application and testing it for variety of load with Apache JMeter load testing tool.

A simple shopping cart application is hosted on AWS in following topologies:

a) *All In One: web server, application server and db server are installed on the same Red Hat Enterprise Linux (RHEL) instance. Refer row number 1 in TABLE 1 below.*
b) *Two tier: Web server and application server are installed on same instance and DB server is installed on a different instance. The web + application server instance is configured to auto-scale [Ref 4] as per the request load on the server. Refer row number 2 in TABLE 1 below.*
c) *Three tier: Web server, application server and DB server are installed on three separate instances. Both the Web server and Application server instances are configured to auto-scale according to the request load. Refer row number 3 in TABLE 1 below.*

Table 1. Web Application Deployment Topologies

Scenario	ELB	Web Server	App Server	DB
1			All in One	
2	Yes	Combined, Auto-Scaled		One Instance
3	Yes	Auto-Scaled	Auto-Scaled	One Instance

The auto scaled environment is configured to work with minimum two instances. This can grow up to twenty instances depending on the load. The software load balancer (HAProxy) is installed on the same server instance which is hosting the web server. A custom component is deployed on each of the HAProxy server instances. This component is responsible for the discovery of newly instantiated application server instances. Once discovered the custom component will register those app server instances with HAProxy server. However there would be only one DB server instance at any given point in time. All the application servers would be pointing to the same DB server instance.

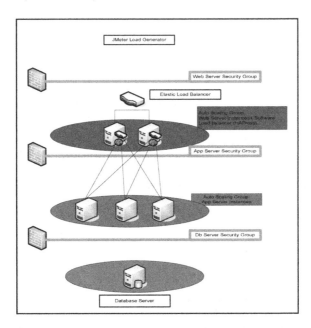

Fig. 2. Web Application Deployment (auto scaling with ELB and HAProxy)

1.4   Components Used

1. *Elastic Load Balancer (ELB)*
    ELB [Ref1] has been used for scenario 2 and scenario 3 as a public interface for web application. It redirects the requests to web server.
2. *Virtual Private Cloud (VPC)*
    VPC [Ref7] is required to create a virtual network subnet which is the key component for setting up the environment for Apache JMeter load testing. This has been used in all of the scenarios.
3. *HAProxy*
    HAProxy [Ref3] is used to balance the load between web server and app server in scenario 3. This is used along with custom component for dynamic node discovery.
4. *Elastic Ip*
    Elastic IP [Ref8] is used in scenario 2 and scenario 3. It is associated with database server and the same is configured in application server.
5. *Auto-Scaling Group*
    Auto-Scaling Group [Ref4] is used for the purpose of application scale out and scale in. This feature is used in conjunction with ELB and HAProxy in scenario 2 and scenario 3.
6. *Security Groups*
    Security groups [Ref 9] act as a virtual firewall. It's a collection of access rules that defines the ingress network traffic to the instance.

2   Learning and Recommendations

Following is the learning outcome from this exercise:

2.1   Load Balancers

ELB does not have a private IP and cannot be wrapped around AWS security group.

- *Description*
  ELB does not have a private IP and security group cannot be wrapped around it,
  which exposes it and makes it open-for-all to access. This behavior does not
  make a sound proposition for inter-server communication between web server
  and app server.
- *Recommendation*
  A software load balancer, like HAProxy [Ref 3], becomes a candidate for load
  balancing the application server layer. These HAProxy instances being normal
  EC2 instances, has private IP addresses and could be protected by security group
  thus making them protected from unauthorized access.

2.2   Auto Discovery of EC2 Nodes

Task is to auto discover the newly instantiated application servers on AWS cloud.

- *Description*
  Application servers being part of auto-scaling group are configured to scale out
  and scale in dynamically. There are no provisions in HAProxy to find out new
  servers and configure itself accordingly to handle new servers.
- *Recommendation*
  To use HAProxy in tandem with auto-scaled application server a custom
  component was developed which discovers the newly instantiated app server
  instance and updates the HAProxy configuration file with app server instance
  information. These auto-scaled application servers, once dynamically registered
  with HAProxy, can be load balanced by the HAProxy instance.

3   Performance Testing Environment

Performance Testing is an important activity of software life cycle. It ensures that the
delivered software meets customer's expectations - varying load, availability and
scalability. In a virtualized environment such as cloud, testing the deployed
application for its performance becomes an important aspect. This section talks about
the observations made during load testing using Apache JMeter performance testing
tool on AWS. It lists out the parameter and corresponding values on which the
comparison of various web application topologies under auto-scaled environment is
done.

This section delineates the steps that were followed to perform load testing. Following points are discussed:

1. *Technical details about software and hardware infrastructure.*
2. *Strategy and parameters used while generating load for application.*
3. *Test results.*
4. *Observations from test results.*

Following diagram (Figure 3) depicts the load testing strategy on AWS. It includes multiple Apache JMeter servers and one UI client.

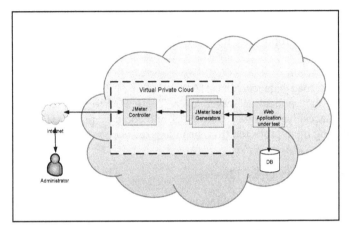

Fig. 3. Apache JMeter setup in Amazon Cloud

3.1   Application Setup

A simple web application was used for performance testing. Following are the details of software and hardware components:

1. *EC2 Instance type [Ref6]– Small (1Elastic Compute Unit, Single Core, 1.7GB RAM)*
2. *Application Server – Tomcat 6*
3. *Database server – MySql 5.5*
4. *Web Server – Apache 2*
5. *Web Application*
6. *Customization done for web application –*
     a) *All application log level is set to FATAL.*
     b) *App server JVM tuning*
7. *Software Load balancer - HAProxy 1.4*
8. *Performance Testing Tool – Apache JMeter*

3.2   Load Generation Strategy

Following points were taken into consideration:
    1. Number of concurrent requests –

a) Scenario 1 (single tier)–
   This scenario was tested by sending 25, 50, 75 and 100 concurrent requests to the application service.
b) Scenario 2 onwards (multi tier)–
   Both these test scenarios were tested by sending 25, 50, 75 and 100 concurrent requests to the application service.
2. Duration of test: 10 minutes.

3.3   Measurable Parameters

Following are the measurable parameters that are taken into consideration:

1. Throughput.
2. Overall Response Time.

3.4   Test Results

Following figure (Fig 4) is a chart with Number of users plotted on X-axis vs. Response time plotted on Y-axis.

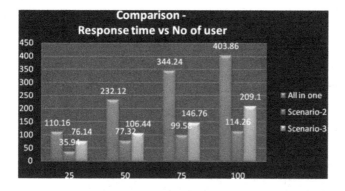

Fig. 4. Number of users vs. response time

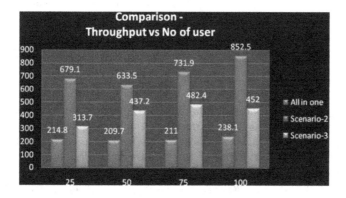

Fig. 5. Number of users vs. Throughput

The chart above clearly shows that the response time is at the minimum for scenario 2 while it is at the maximum in case of scenario 1.

Following figure (Fig 5) is a chart with Number of users plotted on X-axis vs. Throughput plotted on Y-axis.

From the chart shown in Figure 5 it is clear that scenario 2 throughput > scenario 3 throughput is the best suitable topology:

scenario 2 throughput > scenario 3 throughput

Thus, keeping Apache web server and application server on the same instance and auto-scaling this server instance is capable of handling more load than the topology where web server and app server are hosted on separate instances.

Table 2. Comparison Chart for 3 Scenarios

Scenario No.	Number of threads	Throughput	Avg. Through-put	Availability	Configura-bility	Fault tolerance	Maintaina-bility	Scalability
1	25	214.8	218.40	No failover capability	Easy	No	Easy	Not Scalable
	50	209.7						
	75	211						
	100	238.1						
2	25	679.1	724.25	Highly Available	Moderate	Yes	Moderate	Scalable
	50	633.5						
	75	731.9						
	100	852.5						
3	25	313.7	421.32	Highly Available	Difficult	Yes	Difficult	Scalable
	50	437.2						
	75	482.4						
	100	452						

3.5   Observations

- ***Scenario 1****: It gives moderate performance as compared to other 2 scenarios. All 3 servers are installed on same instance causing the degradation in performance and hence the response time. Though the configuration and maintenance of this setup is very easy, it is devoid of any failover capability. At any given point in time there would only be one sever which would be serving all the requests to the application and as the number of requests increases the performance of an application would go on decreasing.*

- *Scenario 2:* From the table above it is clear that Scenario 2 gives the best throughput. Reason being; the database server is kept on separate instance which helps reduce the load on single server, while the web + application server instance handles the request load, which in turn increased the request handling capacity of a server. The availability for this scenario is better than scenario one. Reason being layer 1 was auto-scaled and registered with ELB. Scenario 2 is easier to configure in a sense that it does not require additional software load-balancer and a custom component to be installed for node discovery.
- *Scenario 3:* It gives reduced throughput as compared to Scenario 2. Though the demarcation of servers can be seen here the reduced throughput is due to the intricacy developed by network traffic between web server instance, application server instance and db server instance. Along with this, the time taken for distribution of load at ELB level and HAProxy level also contributes to the cause. Scenario 3 is high on availability due to auto-scaled layer 2 and 3. Software load-balancer and custom component makes configuration and administration difficult as compared to scenario 1 and 2.

3.6   Recommendation

For different type of web applications, the three scenarios that have been documented in this whitepaper may serve as a benchmark. Here is a list of recommendation:

Table 3. Scenario Recommendations

Application Scenarios	Recommended Scenario	Examples
High performance	Scenario 2	Search engines, Data reporting
High availability	Scenario 3	E-commerce, High data streaming applications
High maintainability	Scenario 1	Organization intranet sites

4   Learning

Following is the learning outcome from this exercise:

- *Network Subnet*
  The task is to create a single network subnet on AWS cloud.
  - *Description*
    Apache JMeter works on RMI protocol. RMI cannot communicate across subnets without a proxy. There is no guarantee that all the instances created on AWS public cloud will be on the same subnet. This creates a road block to use JMeter.

o   *Recommendation*
Using Virtual Private Cloud [Ref7] in AWS makes sure that all the instances created within the VPC have the same subnet. JMeter load generators and controllers are placed inside the VPC so that they can generate the distributed load for web application on AWS cloud.

5   Conclusion

Hosting a web application on cloud does addresses the issues involved with traditional data centre, such as hardware provisioning, software installations and infrastructure maintenance. There were certain challenges that were faced while setting up an environment which were solved either by using software or by developing custom components. The hurdle around security with Elastic Load Balancers could be tackled by implementing custom logic along with using software load balancers.

Creating a network subnet which is an integral part for Apache JMeter configuration can be handled by using AWS VPC. As far as the performance of an application is concerned, web application deployed with two tier topology gives optimum throughput on AWS cloud when compared with single tier and 3 tier setup.

Unless and until there is a specific requirement to go for three tier setup, two tier setup is optimal solution. It's fault-tolerant, highly available, easy to configure and maintain, and scalable. Scenario 3 is difficult to configure, due to additional overhead of installing HAProxy and custom component for node discovery. It also lags in performance due to network traffic and time taken by request to route thorough HAProxy.

References and Further Reading

[1] http://aws.amazon.com/elasticloadbalancing/
[2] http://aws.amazon.com/ec2/
[3] http://haproxy.1wt.eu/
[4] http://aws.amazon.com/autoscaling/
[5] http://jakarta.apache.org/jmeter/usermanual/remote-test.html
[6] http://aws.amazon.com/ec2/instance-types/
[7] http://aws.amazon.com/vpc/
[8] http://aws.amazon.com/articles/1346
[9] http://docs.amazonwebservices.com/AWSEC2/2007-08-29/
    DeveloperGuide/distributed-firewall-concepts.html

A Rule-Based Approach for Effective Resource Provisioning in Hybrid Cloud Environment

Rajkamal Kaur Grewal and Pushpendra Kumar Pateriya

Lovely Professional University
Phagwara, India
{grewalrajkamal86,pushpendra.mnnit}@gmail.com

Abstract. Resource provisioning is important issue in cloud computing and in the environment of heterogeneous clouds. The private cloud with confidentiality data configure according to users need. But the scalability of the private cloud limited. If the resources private clouds are busy in fulfilling other requests then new request cannot be fulfilled. The new requests are kept in waiting queue to process later. It take lot of delay to fulfill these requests and costly. In this paper Rule Based Resource Manager proposed for the Hybrid environment, which increase the scalability of private cloud on-demand and reduce the cost. Also set the time for public cloud and private cloud to fulfill the request and provide the services in time. The Evaluated the performance of Resource Manager on the basis of resource utilization and cost in hybrid cloud environment.

Keywords: Cloud computing, Provisioning, VIM, Hybrid cloud, IaaS, Resource Allocation.

1   Introduction

Cloud computing refers technology that enables functionality of an IT Infrastructure, IT platform or an IT product to be exposed as a set of services in a seamlessly scalable model so that the consumers of these services can use what they really want and pay for only those services that they use(Pay per use). The agency can host a cloud itself, the subscribe to the cloud service from another third-party service provider. The Office of management and the budget, National Institute of Standards and Technology are describing standards for cloud procurement and acquisition Cloud computing embraced by small medium businesses. Large enterprises have started using for select workloads and business need.

Cloud computing is about moving services, computation or data–for cost and business advantage-off–site to an internal or external, location transparent, centralized facility or contractor. By making data available in the cloud, it can be more easily and ubiquitously accessed, often at much lower cost, increasing its value by enabling opportunities for enhanced collaboration, integration and analysis on a shared common platform [1].

S. Patnaik et al. (Eds.): *New Paradigms in Internet Computing*, AISC 203, pp. 41–57.
DOI: 10.1007/978-3-642-35461-8_5   © Springer-Verlag Berlin Heidelberg 2013

The definition of cloud computing as per Gartner is "A style of computing where massively scalable IT-enable capabilities are delivered as a service to external customers using internal technologies". Cloud computing is delivery of computing as a services rather than product. Cloud 'Services' refer to those types of services that are exposed by cloud trafficker and that can be used by cloud consumer on a 'pay-per-use' basis. An ongoing monthly expenses easy to incorporate into budget than large one-time outlay. In this cancel and change subscription at any time without losing large initial investment.

In practice, cloud services are classified–Infrastructure as a Services(IaaS), Software as a Services(SaaS), Platform as a Services(PaaS), Data as a Services (DaaS)[2].

In Infrastructure as a service large amount computing resources are managed by IaaS provider which is allocated to user as demanded. IaaS is the concept of providing Hardware as a Service. IaaS offer basic storage and compute capabilities as standardized services over the network. Servers, storage systems, switches, routers, and other systems are pooled and made available to handle workloads that range from application components to high-performance computing application [2]. There are a number of successful IaaS providers: Amazon EC2, Joyent, Rack space etc.

Platform as a Service is a way to rent hardware, operating systems, storage and network capacity over the internet. It delivers a computing platform or software stack as a service to run applications. This can broadly be defined as application development environment offered as a 'service' by the vendors. The development community can use these platforms to code their applications and then deploy the applications on the infrastructure provided by the cloud vendor. Here again, the responsibility of hosting and managing the required infrastructure will be with the cloud vendor. AppEngine, Bungee Connect, LongJump, Force.com, WaveMaker are all instances of PaaS.

Software as a Service also refers to application as a Service. Instead of running as application locally, the application resides on the cloud and online alternatives such as access via web interface is provided. SaaS is the service based on the concept of renting software from a service provider rather than buying it yourself. That eliminates the need to install. Yahoo mail, Google docs applications, Salesforce.com CRM apps, Microsoft Exchange Online, Facebook are all instances of SaaS.

Data as a Service is refer data in various formats and from various sources could be accessed via services by users on network, in a transparent, logical or semantic way. Users could, for example, manipulate remote data just like operate on a local disk or access data in a semantic way in the Internet. The DaaS could be found at some popular IT services, e.g. Google Docs and Adobe Buzzword. A hybrid cloud is a composition of at least one private cloud and at least one public cloud. In Quality of services guaranteed the computing Clouds can guarantee QOS for users and Service Level Agreement is negotiated between cloud provider and user on the level of availability, serviceability, performance, operation, or other attributes of the service like billing and even penalties in the case of violation of the SLA[6].

Cloud computing provide services from hardware to application. The services are-Infrastructure as a Service(IaaS), Software as a Services(SaaS) and Platform as a Service(PaaS). Proposed methodology is based on Infrastructure as a Service layer for receive resource on demand. In this can choice of virtual computer, meaning the can select a configuration of CPU, memory and storage that is optimal for your application. The IaaS provider supplies the whole cloud infrastructure viz. servers, routers, hardware based load-balancing, firewalls, storage and other network equipment. The customer buys resources as a service on an as needed basis.

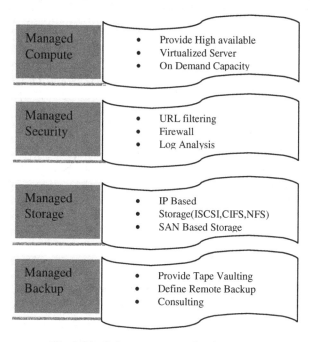

Fig. 1. The Infrastructure as a Service – Offer

The Spectrum of Cloud

Table 1. The spectrum of cloud

IaaS	PaaS	SaaS
Amazon	AppEngine	Google Apps
Rackspace	Heroku	Taleo Workday
Linode	Force.com	Salesforce
GoGrid	Azure	
And Other	Heroku	

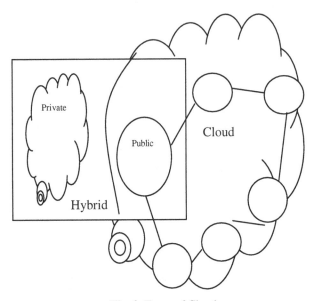

Fig. 2. Types of Cloud

Public cloud, Private cloud and hybrid cloud are types of cloud computing.
Public cloud, Private cloud and hybrid cloud are types of cloud computing.

- The Public clouds are run by third parties and the applications from different customers are too mixed together on the cloud's servers, storage systems, and networks. The datacenter hardware and software is what we will call a Cloud. It is external services clouds based on pay-per-use model. The Public cloud or external cloud define cloud computing with traditional mainstream where resources are provisioned on self-services basis over the internet.
- The Private clouds are exclusively built and managed by a company's own IT organization or by a cloud provider. We use the term Private Cloud to refer to internal datacenters of a business or other organization, not made available to the general public. The benefits with private cloud are higher security and flexibility to manage the cloud according to the need.
- The hybrid cloud environment consisting multiple internal and external providers. The Hybrid clouds combination of both public and private clouds. It can help to provide on-demand, externally provisioned scale, higher security and efficiency. Hybrid Clouds provide more secure control of data and applications and that allow various parties to access information over the Internet.

The Differences between Cloud and Grid Computing
The Grid computing is distributed computing and parallel computing where a super and virtual computer is composed of a clusterof computer to perform tasks. That motivated by real problems appearing in advanced scientific research. Cloud computing provides on-demand resource provisioning.

- The Cloud computing is directly pulled by immediate user needs driven by various business requirements. The Grid defines re-usability for high performance computing.
- The Cloud owned by a single physical organization, who allocates resources to different running instances. The Grid defines the resource sharing to form the virtual organization.
- Grid is better for applications that are large and need large amount of computing power. Whereas cloud computing is recommended when there are large amount of applications asking for comparatively less computing power.
- The Cloud and Grid both provide data storage. The Amazon S3 is a cloud storage service provider in which data objects containing 1 byte to 5 GB of data can be stored. Storage in the data grid is efficient for data-intensive storage, it is not efficient for storing 1 byte small objects.

Why Cloud Computing Important

Provide Scalability on Demand

All the organization change according in environment. If any organization has some period of time in which computing resource needs much higher than cloud can deal with this changes.

Improving the Business Processes

An Organization and its suppliers and partner can share application and data in the cloud and that allow everyone involved the focus on the business process. The cloud provides an infrastructure for business process improving.

Minimize Costs

Organizations start with infrastructure so the time and any other resources that would spent on building data center borne by cloud provider. The cloud computing reduce startup costs.

Streamlining the Data Center

Any size of organization will have investment in data center. That also buying and maintaining the software and hardware who keep the data center running. But the organization can streamline data center by taking cloud technologies internally and reduce workload into public.

Provide High Scalability

The "Cloud" the size of dynamically scalable to meet the growing size of applications and user needs.

On-Demand Service

The "Cloud" provide on-demand services.Cloud is huge resource pool. We can purchase it on demand.

Versatility

The cloud computing is not for particular application.Cloud can support different applications running simultaneously.

Virtualization

In Cloud computing the users users at any location,using the variety of terminal acces to application services.The application in cloud somewhere in the running but the user don't know about the specific location of running application.

2   Objectives

To design an efficient rule base approach for resource provisioning in hybrid cloud environment. This will satisfy some basic objectives:

- Proposed method for satisfy a minimum time-criticality.
- Provide resources at minimum cost.
- Support scalability.
- Priorities the requests on the basis of resource requirement.
- Give the optimal throughput (requests/time unit).
- The main objective in cloud computing to improved resource utilization by sharing available resources to satisfy multiple requests effectively and efficiently.
- Provide the services according user need within the time without waste time for waiting.

3   Resource Manager

Cloud computing is services based resources. In cloud environments, the customer able to access only the services they entitled to use. The Entire applications may be used on a cloud services basis. The Provisioning is the process of preparing and a network to allow to providing services to users. The Resource management is the efficient and effective deployment of an organization's resources when they are needed. Eucalyptus, Hadoop and Open Nebula are open source cloud computing framework. It has its own architecture.

Resource management system in data centers are support Service Level Agreement-oriented resource allocation. The SLA Resource Allocator acts as the interface between the Cloud computing infrastructure and external users/broker. The user interact with the Cloud management systems by an automatic system like broker who act on users behalf for the submit service requests from anywhere in the world to the Clouds to be proposed. The resource provisioning stage is the time when the cloud broker make decision to provisioning resource by on-demand plans it allocate VMs to cloud providers for utilizing the resources. The provisioning stages based on number of planning epochs that considered by cloud broker.

In these models all the requests of users for resources handle through cloud services manager. In this user request for the resources that handled through cloud service manager, which manage all resources and billing. Next CSM interact with Virtual Infrastructure Manager for user's resource requests. A user requests for the resources. Virtual machine next interacts with the Data Center Broker to fulfill request from available existing resources. The Cloud brokers to serve as intermediaries between end users and cloud providers. The multiple VMs can concurrently run applications based on different operating system environments on single physical machine .The multiple Virtual machines started and stopped dynamically to meet accepted services requests.

Physical hosts, data center broker and VIM form a cloud. That is key element of cloud at Infrastructure as a services layer. This layer also called Hardware as a service layer. All the services of cloud are provides by cloud services manager. When user requests for resources the resource manager accept user request and tries to fulfill of user request for resources on private cloud. More preference of its resource utilize are given by private clouds. If the resources that are user requested are available in private cloud then request is fulfilled by allocating these resources. If the resources of private cloud are assign with other application, then resource manager allocate resource resources of public cloud by interact with its CSM. If the resources of private cloud are busy to fulfill other requests, then new request are kept in waiting queue to process later when that resource are available. It makes delay for provisioning resource and scalability is limited of private cloud.

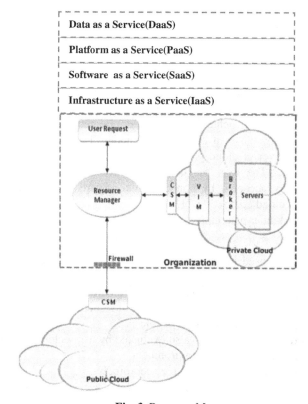

Fig. 3. Resource Manager

With Rule Based Resource Manager a private cloud can be scaled up to allocated resources on-demand even if that is overloaded. Rule Based Resource Manager proves to cost effective in term of utilization of private cloud resources.

Cloud Computing is service-based resources. In cloud environments the customer is able to access only the services that they need and use. Resource Manager is used for resource provisioning in hybrid environment. To increase scalability, the resources are borrowed from public cloud. The important issue is efficiently managing the allocation of resource needs between two clouds. The efficiency is determined by resources utilization and cost for using public cloud.

4   Related Work

From the last fewer, cloud computing has evolved as delivering software and hardware services over the internet. The extensive research is going on to extend the capabilities of cloud computing. Given below present related work in the area of cloud's scalability and resource provisioning in cloud computing.

In 2010 ChunyeGong, Jie Liu, Oiang Zhang, Haitao Chen and Zhenghu has discussed Characteristics of Cloud Computing. In this paper summarize the general characteristics of cloud computing which will help the development and adoption of this rapidly evolving technology. The key characteristics of cloud computing are low cost, high reliability, high scalability, security. To make clear and essential of cloud computing, proposes the characteristics of this area which make the cloud computing being cloud computing and distinguish it from other research area.

The cloud computing has its own technical, economic, user experience characteristics. The service oriented, loose coupling, strong fault tolerant, business model and ease use are main characteristics of cloud computing. Abstraction and accessibility are two keys to achieve the service oriented conception. In loose coupling cloud computing run in a client-server model. The client or cloud users connect loosely with server or cloud providers. Strong fault tolerant stand for main technical characteristics. The ease use user experience characteristic helps cloud computing being widely accepted by non computer experts. These characteristics expose the essential of cloud computing. [1]

In 2010 Pushpendra kumar pateria, Neha Marria discussed resource provisioning in sky environment.

- Resource manager is used for resource provisioning and allocate of resource as user request.
- Offer the rule based resource manager in sky environment for utilization the private cloud resource and security requirement of resource of critical application and data. Decision is made on the basis of rule.
- Performance of resource manager is also evaluated by using cloudsim on basis of resource utilization and cost in sky environment.
- Set priorities request and allocate resource accordingly.
- Sky computing provides computing concurrent access to multiple cloud according user requirement.

- Define the Cloud services like Software as a service (SaaS), Platform as a Service(PaaS) and Infrastructure as a service (IaaS). Also define the architecture of cloud computing.[2]

In 2010 Zhang Yu Hua, Zhang Jian ,Zhang Wei Hua present argumentation about the intelligent cloud computing system and Data warehouse that record the inside and outside data of Cloud Computing System for data analysis and data mining. Management problem of CCS are: balance between capacity and demand, capacity development planning, performance optimization, system safety management. Architecture of the Intelligence cloud computing system is defined with Data source, data warehouse and Cloud computing management information system. [3]

In 2008 discussed about the Phoenix by Jianfeng Zhan, Lei Wang, Bipo Tu, Yong Li, Peng Wang, Wei Zhou and Dan Meng. In this paper discuss the designed and implemented cloud management system software Phoenix Cloud. Different department of large organization often maintain dedicate cluster system for different computing loads. The department from big organization have operated cluster system with independent administration staffs and found many problems. So here designed and implemented cloud management system software Phoenix Cloud to consolidate high performance computing jobs and Web service application on shared cluster system. Also imposed cooperative resources provisioning and management policies of large organization and their departments to share the consolidated cluster systems. Phoenix Cloud decreases the scale of required cluster system for a large organization, improve the benefit of scientific computing department, and provisions resources [4].

In 2010 Shu-Ching Wang, Kuo-Qin Yan, Wen–Pin Liao and Shun-Sheng Wang discussed about Load Balancing in Three-Level Cloud Computing Network. Cloud computing use low power host to achieve high reliability. In this Cloud computing is to utilize the computing resources on the network to facilitate the execution of complicated tasks that require large-scale computation. In this paper use the OLB scheduling algorithm is used to attempt each node keep busy and goal of load balance.

Proposed LBMM scheduling algorithm can make the minimum execution time of each task on cloud computing environment and this will improve the load unbalance of the Min-Min. In order to reach load balance and decrease execution time for each node in the three-level cloud computing network, the OLB and LBMM scheduling algorithm are integrated. The agent collects related information of each node participating in this cloud computing system. In the proposed Method, services manager that passes the "threshold of services manager" considered effective and will be the candidate of effective nodes by manager. The Threshold of service node is used choose the better service node. The load balancing of three-level cloud computing network is utilized, all calculating result could be integrated first by the second-level node before sending back to the management[5].

In resource provisioning resource are allocated to applications with service level agreement (SLA). The Ye Hu, Johnny Wong, Gabriel Iszlai and Marin Litoiu discussed resource allocation to an application mix is done such that SLA of all application is met. Here two server strategies namely shared allocation (SA) and dedicated allocation (DA) are considered for the resource allocation. The allocation strategies evaluated by heuristic algorithm on basis of smallest number of servers required to meet the negotiated SLA. Probability dependent priority found. [6]

In 2011 M.Noureddine and R.Bashroush demonstrate modality cost analysis based methodology for cost effective datacenter capacity planning in the cloud. Provisioning appropriate resources to each tenant /application such that services level agreements (SLA) met. The objectives of this paper to use methodology to guide resource provisioning. The systematic methodology to estimate the performance of each modality. The quantitative methodology explained for planning the capacity of cloud datacenter. Set the applications in modalities and measure the cost of hardware resources. Discussed three experiments,MCA-S, MCA-M and MCA-L that represented user profiles and measure the resource overhead. A validation tool is used to simulate load and validate assumptions. Office Lync Server Stress (LSS) generate simulated load.[7]

In January 31, 2011, Sivadon Chaisiri, Bu-Sung Lee, and Dusit Niyato discuss about the Optimization of Resource Provisioning Cost. Under the resource provisioning optimal cloud provisioning algorithm illustrates virtual machine management that consider multiple provisioning stages with demand price uncertainty. In this task system model of cloud computing environment has been thoroughly explained using various techniques such as cloud consumer, virtual machine and cloud broker in details. [8]

The agent-based adaptive resource allocation is discussed in 2011 by the Gihun Jung, Kwang Mong Sim. In this paper the provider needs to allocate each consumer request to an appropriate data center among the distributed data centers because these consumers can satisfy with the service in terms of fast allocation time and execution response time. Service provider offers their resources under the infrastructure as a service model. For IaaS the service provider delivers its resources at the request of consumers in the form of VMs. To find an appropriate data center for the consumer request, propose an adaptive resource allocation model considers both the geographical distance between the location of consumer and data centers and the workload of data center. With experiment the adaptive resource allocation model shows higher performance. An agent based test bed designed and implemented to demonstrate the proposed adaptive resource allocation model. The test bed implemented using JAVA with JADE (Java Agent Development framework). [9]

Dongwan Shin and Haken Akkan discussed about the "Domain based Virtual Resource Management in Cloud Computing". Cloud Computing enable convenient, on-demand access to computing resource.

- For satisfying various needs from user of different groups, cloud computing provides benefits from computer security. The critical challenge in this computing demanding advnced mechanism for protecting data and applications in private, public and hybrid cloud, so for instance they include cloud data security in clouds.
- In this paper motivated to investigate a flexible, decentralized, and policy-driven approach to protecting virtualized resources.

- The delivery model of cloud computing called infrastructure as a service (IaaS) provides users with infrastructure services that depend upon virtualization techniques. IaaS services provide have a user-based service model. With increase popularity of cloud computing, there demand for more scalable and flexible IaaS model with fine gained access control mechanism.

- In this paper for provisioning and managing users and virtualized resources in IaaS propose domain-based framework. Cloud service provider delegates its functions to each domain. Domains manage their users and virtualized resources assignee from cloud services provider using role based security policy.

- Role based security policy demonstrate in computer security communities. The role-based access control (RBAC) use intermediary concept called role for provide an indirection mechanism between users and permissions.

- Domain based framework provides scalable user resource management, domain-based and flexible pricing.

- To support the services cloud computing consist of three components depending on its deployment models. That are 1)hardware and facilities 2)software kernel and 3)virtualization that classify the services and application of cloud computing like computing and storage resource) cloud software development platform, and 3)cloud software application.

- Finally Discussed about how to design and implement proof-of-concept prototype by modifying existing Eucalyptus that has no notion of roles for its access control.[9]

In 2010 Jianfeng Zhana, Lei Wanga,Weisong Shib, Shmin Gonga,Xiutao Zanga discussed about Resource Provisioning. Resources providers needs to provision resources for heterogeneous workloads in different cloud scenarios as more and more services providers choose cloud platforms which is provided by third party resource providers. They have focused on some important issue.

- Propose coordinated resource provisioning solution for heterogeneous workloads in two typical cloud scenarios. First is a large organization operates a private cloud for two heterogeneous workloads .Second, a large organization or two service providers running heterogeneous workloads revert to a public cloud.

- Build agile system, called Phoenix Cloud that enables resource provider to create coordinated runtime environments on demand for heterogeneous workloads when they are consolidated on the cloud site.

The Comprehensive evaluation has been performed in experiments. Typical heterogeneous workload traces: parallel   batch jobs and web services, the experiments show that:

o In a private cloud scenario, when the throughput is almost same like that of a dedicated cluster system, solution decreases the configuration size of a cluster by about 40%.

o In public cloud scenario, solution decreases not only the total resource consumption but also the peak resource consumption maximally to 31% with respect to that of EC2 plus right scale solution.

5   Rule Based Resource Manager

Hybrid cloud is a composition of two or more clouds that remain unique entities but that are bound together, offering benefits of multiple deployment models. In hybrid cloud we propose Rule Based Resources Manager for successfully utilizing the private cloud resources and considering the security requirements of applications and data. With resource manager a private cloud can be scaled up to allocate resources on-demand even if private cloud overloaded. Also the scalability beyond the capacity of private cloud is achieved by using public cloud resources. Decision is made on the basis of rules presented in following paragraph.

As shown in figure 1, User request for resources and its request enter to the Resource Manager. In our approach we categorized user's request into two types based on their resource requirements that is critical data processing and data security. These requests are assigned according to priority, if the user's need to perform critical data processing or security demand is high then the request is classified as a high priority and if the request is to run non critical tasks then it classified as low priority. The Resource Manager recognizes the suitable cloud to be used to fulfill a request. The high priority request always access resources from the private cloud itself, because it have confidential (secure) information. Next low priority requests can be fulfilled from either public cloud or private cloud. But if the private cloud resources are available, it must be used first as these resources are possess by the enterprise and should be utilized. Sometimes high priority request fulfilled by private cloud but its resources are already assigned to fulfill previous requests of low and high priority. In this situation we find those already allocated low priority requests and reallocate these low priority requests for which the remaining cost on public cloud is minimum to public cloud. The algorithm for Resource Manager is as follows:

VM = Virtual Machine

New_VM_Requestc+s = New Virtual Machine Request for Compute and Storage Resources

Availablec+s = Available Compute and Storage Resources in Private Cloud

High_Priority_Request= Request which Performs Critical Data Processing and Needs Good Response Time

Low_Priority_Request = Request which does not Perform Critical Data Processing

x = Allocated Low Priority Virtual Machines

xc+s = Compute and Storage Resources of x

Relocation_Cost_on_Public_Cloud = Public Cloud Usage Cost of a VM, when It is Reallocated from Private Cloud to Public Cloud.

Waiting_Queue = VM Requests are Put in Waiting Queue for Allocation on Private Cloud When Its Resources become Available.

RESOURCE_MANAGER

(New_VM_Requestc+s)
{

PROCEDURE 1

If (New_VM_Requestc+s <= Availablec+s)

{

    Then:

/*--------This Redirects New_VM_Requestc+s to Private_Cloud -----*/

  Allocate NEW_VM Requestc+s on Private Cloud

    Response.Redirect (Private_Cloud);

}

/*--- Rule Two will always ensure high priority Requests on Private server

PROCEDURE 2

If (New_VM_Requestc+s > Availablec+s &&
New_VM_Requestc+s =High_Priority_Request)

{

/*(For First 10milli-sec check low priority on private server) */

For Time 0 to 10ms

{

/* -- It Will free Space from Private Cloud Sever---*/

Check Low Priority Request on Private sever

}

If (Low Priority.Request.Count<0)

{

 Put New _VM_ Requestc+s in Waiting_ Queue

[Because no Space on private server so waiting Queue]

If (time>100ms)

{

  Move To another public server

  Break;

}

Else

If (Low Priority.Request.Count>0)

{

Re-Allocate Existing Low_Priority_Request to   Public Sever

Then

Handle New_VM_Requestc+s to Private Server

 }

If (time >100 ms and less than 200ms)

{

   Put New_VM_Requestc+s in Waiting_Queue

}

If (time>=200ms)

{

   Move To another public server

}

}

PROCEDURE 3

/*--------This Redirects to Public Cloud-----*/

If (New_VM_Requestc+s > Availablec+s &&
New_VM_Requestc+s ==
Low_Priority_Request)

{

    Redirect Requests to Public _Cloud;

}

6   Experimental Evaluation

With performance of Resources Manager experimental results are present. CloudSim3.0 is used for simulation. CloudSim is a toolkit of simulation of Cloud Computing Scenarios. It provides basic classes for describing data centers, virtual machines, applications, users, computational resources, and policies for management of diverse parts of the system. CloudSim is a simulation framework which supports seamless modeling and experimentation of cloud computing infrastructure, including datacenters on a single computer. It has a virtualization engine, which assists in creation and management of multiple, independent, and co-hosted virtualized services on a data center node. It supports performance evaluation of policies for resource provisioning and application scheduling. CloudSim written in java 6.1. Java is a general-purpose, concurrent, class-based, object-oriented language. Java is a programming language designed for use in the distributed environment of the internet.

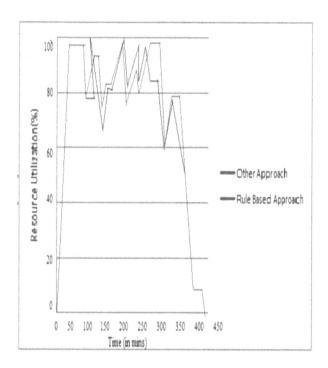

Fig. 4. Utilization Resources of Private Cloud

The private and public cloud having 20 and 90 heterogeneous type servers are considered for simulation. Pay-per-use model is used for billing. Few requests are getting when private cloud is overloaded, we made 50 resource allocation requests at different time intervals that is–low priority and high priority. Set the time for public cloud and private cloud to fulfill the request. For Rule Based and Non-Rule Based approaches we simulated different combinations of priorities among these requests. Computing resource is more costly than storage resource utilization. We consider only computing resource utilization. Private cloud utilization is found to be 64% in Non Rule Based approach and 70.59% for Rule Based approach. Our approach depends on Rule Based Resource Manager that proves to be cost effective in money spent for using cloud resource. Our approach considers security requirements of confidential and never permits to pass the organization's firewall and provide the services in time.

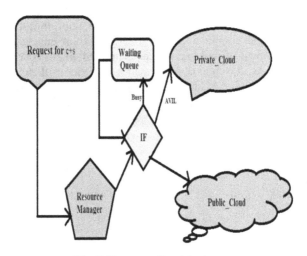

Fig. 5. Resources Provisioning

Table 2. Comparison of Resource Utilization Under Different Set of Priority

Resource Utilization	Resource Utilization in Rule Based Approach	Resource Utilization in Non-Rule Based Approach
High Priority Requests	70.38%	65.73%
Low Priority Request	64.70%	65.73%
Mixed Priority Request	70.10%	65.73%

7 Conclusion

Under the resource provisioning in Cloud Computing, the long-held dream of computing as a utility, has the potential to transform a large part of the IT industry, making software even more attractive as a service and shaping the way IT hardware is designed and purchased. Developers with innovative ideas for new Internet services no longer require the large capital outlays in hardware to deploy their service or the human expense to operate it. They need not be concerned about over- provisioning for a service whose popularity does not meet their predictions, thus wasting costly resources, or under- provisioning for one that becomes wildly popular, thus missing potential customers and revenue. The Methodology based on Infrastructure as a Service layer to access resources on-demand. A Rule Based Resource Manager is proposed to scale up private cloud and presents a cost effective solution in terms of money spent to scale up private cloud on-demand by taking public cloud's resources and that never permits secure information to cross the organization's firewall in hybrid cloud. Also set the time for public cloud and private cloud to fulfill the request.

References

[1] Gong, C., Liu, J., Zhang, O., Chen, H., Gong, Z.: The Characteristics of Cloud Computing. In: Parallel Processing Workshops, pp. 275–279 (2010)

[2] Pateria, P.K., Marria, N.: On-Demand Resource Provisioning in Sky Environment. International Journal of Computer Science and its Application, 275–280 (2010)

[3] Hua, Z.Y., Jian, Z., Hua, Z.W.: Discussion of Intelligent Cloud Computing System. In: International Conference on Web Information Systems and Mining, pp. 319–322 (2010)

[4] Zhan, J., Wang, L., Tu, B., Li, Y., Wang, P., Zhou, W., Meng, D.: Phoenix Cloud: Consolidating Different Computing Loads on Shared Cluster System for Large Organization. In: Proceeding of the First Workshop of Cloud Computing and its Application (July 17, 2010)

[5] Wang, S.-C., Yan, K.-Q., Liao, W.-P., Wang, S.-S.: Towards a Load Balancing in a Three-level Cloud Computing Network

[6] Hu, Wong, J., Iszlai, G., Litoiu, M.: Resource Provisioning for Cloud Computing. In: Conference of Center for Advanced Studies on Collaborative Research (2009)

[7] Noureddine, M., Bashroush, R.: Modality cost analysis based methodology for cost effective datacenter capacity planning in the cloud. Special Issue on the International Conference on Information and Communication System, ICIC 2011, pp. 1–9 (2011)

[8] Chaisiri, S., Lee, B.-S., Niyato, D.: Optimization of Resource Provisioning Cost in Cloud Computing, pp. 1–32

[9] Jung, G., Sim, K.M.: Agent-based Adaptive Resource Allocation on the Cloud Computing Environment. In: 2011 40th International Conference on Parallel Processing Workshops (ICPPW), pp. 345–351 (2011)

[10] Suhail Rehman, M., Sakr, M.F.: Initial Findings for Provisioning Variation in Cloud Computing. In: International Conference on Cloud Computing, pp. 473–479 (2010)

[11] Shin, D., Akkan, H.: Domain based Virtualized Resource Management in Cloud Computing

[12] Goudarzi, H., Pedram, M.: Maximizing profit in cloud computing system via resource allocation, pp. 1–6. University of Southern California, Los Angeles (2011)

Browser Based IDE to Code in the Cloud

Lakshmi M. Gadhikar, Lavanya Mohan, Megha Chaudhari,
Pratik Sawant, and Yogesh Bhusara

Fr. C.R.I.T,
Vashi,
Navi Mumbai
{lavanyam.210,megha.chaudhari}@gmail.com
yogeshbhusara@hotmail.com

Abstract. Cloud computing is the future of computing. What makes it stand apart is the fact that everything including data and applications are stored on the cloud itself and are accessible through an Internet connection and a web browser. Integrated Development Environment (IDE) which is used for coding and developing applications can be a software on the cloud. The aim of this paper is to convey the idea of Cloud based IDE for the Java language which will have the features to write, compile, run and test code on the cloud. This software will permit easy development of Java applications. It will provide sharing of code and real time collaboration with peers. It will also have an integrated forum and a technical blog. This can be used by the developers who require instant help related to coding Java applications. This software can be used instead of or alongside a desktop IDE. The only requirement for accessing and using this application is a web browser and an Internet connection. This eliminates hardware and operating system issues thus allowing people with different hardware and heterogeneous operating systems to collaborate and code with ease. It also eliminates the need to use conventional devices like desktops or laptops to code programs by allowing the users to access this IDE from various devices like smart phones that have an Internet connection.

Keywords: Cloud Computing, Integrated Development Environment (IDE), Browser Based IDE.

1 Introduction

The latest trend is to take desktop applications online and to provide them as a service. There are already many online editors like Google Docs or Zoho Writer. Just like these text editors, even Integrated Development Environments can be taken online and provided as a service.

Today, people are facing an increasing need to program from anywhere, anytime. Also, a need is felt to be able to program from devices other than just the desktops and laptops. It would also be better if there was no requirement of downloading and installing software for the purpose of coding and running of programs e.g. the JDK or

S. Patnaik et al. (Eds.): *New Paradigms in Internet Computing,* AISC 203, pp. 59–69.
DOI: 10.1007/978-3-642-35461-8_6   © Springer-Verlag Berlin Heidelberg 2013

a desktop IDE. Also, today, different people across the globe are involved in the same projects. So, new capabilities of sharing and collaboration are required.

All these requirements can be fulfilled with the help of an online IDE. An online IDE which can also be called a browser based IDE is an online coding environment that allows users from across the globe to access and use this software as a service. This software can be accessed from devices like smart phones, desktops, laptops, etc. that have the ability to access the Internet and have a web browser.

These browser based IDEs can be hosted on a cloud or as a normal web hosting. In this paper, we present the idea of a browser based IDE on the cloud. The main reason for choosing cloud hosting over client-server architecture are the advantages of cloud hosting like virtually limitless computing power, scalability, risk protection etc. In the classic hosting, there will be only a single physical machine and all the servers, etc. will have to be leased out or bought in advance. But, with cloud computing, there is no large initial investment like in the traditional method and one has to pay only as per his usage.

The users may want to share their projects with their peers and may also want to work on certain modules simultaneously. This can be achieved by the feature of real time collaboration that will be provided by this IDE.

Also, when the developer faces with certain problems during implementation, he may find the need to ask other programmers for help to solve his problems. This need can be catered to by having an integrated forum in the IDE. This will help the developers to obtain answers to their queries.

Some developers may want to share their technical knowledge so that it can be of help to the other developers. For example, a developer might have found out a more efficient way of doing a certain task and may want to share this knowledge with other developers. He will be able to do so conveniently with the help of technical blogs.

In this paper, we present the idea of browser based IDE on the cloud that will have an integrated forum to assist the developers and will also allow the developers to write technical blogs to share knowledge as well as to make certain new findings known to others.

1.1   Cloud Computing

Cloud computing is an Internet based computing which aims at providing hardware and software resources. It enables the users to access and share information from devices like laptops, desktops, smart phones, etc. which have ability to connect to the Internet. Cloud computing caters to dynamism, abstraction and resource sharing [1].

Dynamism deals with the fluctuating demands based on seasonal traffic burst, world or regional economy, etc. Abstraction allows developers to concentrate on core competencies and eliminate the need to worry about operating system, software platform, web security, updates, etc. Resource sharing provides flexibility to share applications and also other network resources like hardware.

Cloud computing allows users from all around the world to access the applications without having to download or install them on their own machines. It can provide virtually unlimited storage as opposed to local servers or hard drives. It provides three major types of services- Infrastructure as a Service (IaaS), Platform as a Service (PaaS) and Software as a Service (SaaS).

1.2   Integrated Development Environment

An Integrated Development Environment (IDE) is a program for software developer that combines the functions of a text editor, an interpreter or a compiler and run time facility to simplify coding and debugging [2]. It has certain important features like syntax highlighting, automatic editing, automatic code completion, compilation, execution, debugging, access database, GUI builder, etc.

1.3   Browser Based Integrated Development Environment

Browser Based IDEs are Integrated Development Environments that are accessible to everyone through a web browser and an Internet connection. It is elevating the coding platform to an online environment where OS issues are eliminated and hardware independence is achieved [3]. It provides increased portability and accessibility. No installation or downloading is required in order to use this. It can be used as a programming environment for multiple people. It is software that is provided as a service.

1.4   Real Time Collaboration

Real Time Collaboration is a technique that allows two or more users to edit the same file using different computers at the same time. This generally uses the Internet and the presence technology for collaboration with the different users. Real time collaboration is useful in applications like distance learning, code review, telephonic interviews, etc.

1.5   Forums

An Internet forum is a discussion area on the website [4]. The members can post discussions or queries. These can be read by other forum members. The other members can respond to these posts.

1.6   Technical Blogs

A blog is a website where one writes things on an ongoing basis [5]. It is like a personal journal online. A technical blog is a blog where people write about IT related technologies and about the happenings in the fast-changing world. These blogs consists of a whole range of topics in the IT world like Internet security, web hosting, new algorithms, etc.

2   Existing Browser Based Coding Environment

A few existing browser based coding environments are Cloud9 IDE, CodeRun Studio, ideone, Eclipse Orion, eXo Cloud IDE, etc.

2.1   Cloud9 IDE

Cloud9 IDE supports HTML, CSS, JavaScript, etc [6]. It is mainly for web development. It has support for real time collaboration.

2.2   CodeRun Studio [7]

CodeRun Studio supports ASP.net, C#.net, Silverlight as well as the languages supported by Cloud9 IDE. It allows sharing of code through hyperlinks.

2.3   Eclipse Orion [8]

Eclipse Orion is mainly for web development and it supports HTML and JavaScript.

All these above IDEs do not have support for Java language.

2.4   Ideone

Ideone is not an IDE. It is like a pastebin that supports compilation and debugging of code in many languages including Java [9]. But it does not permit the creation of projects.

2.5   eXo Cloud IDE

eXo Cloud IDE is the only cloud-based IDE that supports programming in Java language [10]. But it does not support real time collaboration.

3   Proposed System

The idea in this paper is to build browser based IDE for Java language to code in the cloud. This IDE will be present on the cloud and will allow any person to code, compile and run Java code from anywhere in the world and with devices like smart phones, laptops, desktops, etc. that have a browser, an Internet connection as shown in Fig. 1.

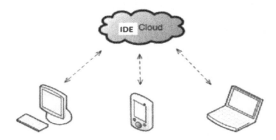

Fig. 1. Browser based IDE on the cloud accessible from various devices

The IDE will also have the feature of sharing the code with peers and real time collaboration. The IDE will also have an integrated forum and blogging facility for the developers.

This cloud application will overcome the limitations of the coding environments on the desktop viz. the need to download and install desktop based IDEs, hardware and operating system issues, etc. This application will overcome existing browser based IDE for Java language by adding new and special functionality of easy sharing of code with peers through real time collaboration. Real time collaboration will be used when there exists a team of developers working on the same files or projects. The users will have access to the same files at the same time and will be permitted to modify the files simultaneously. The edits will be viewed by all the developers at run time. This functionality will also be useful in telephonic interviews where the interviewer can assess the candidate by asking him to write certain code and viewing the same code in real time.

This IDE will also support a forum where the developers can ask queries and obtain answers or solutions to their problems. This will add to the convenience of the developers as they will not have the need to login to other online forums separately in order to ask questions and obtain answers. This forum will also have an additional feature of access modifier. The developers who ask questions can mark it as either private or public. The public questions will be available to all the IDE users and anyone can answer it. But, there may be cases where the developer is having problems with certain portion of the code. He may want to post certain snippets of the code online so that others can help. But, at the same time, he may not want the code to be visible to all. In such cases, the developer can mark the question as private and specify the name of the project that the question deals with. On doing so, this question will be available to only those developers with whom the project has been shared. This feature will be useful for security purposes.

The IDE will also have support for technical blogs. The users of the IDE can have their own personal blogs where they can share technical information. Every blog post will have to be tagged. The post can have tags like 'swings', 'hibernate', 'algorithm', etc. This will be useful when the developer wants to search contents related to certain topics. This tag will also be helpful in sending emails to the developers in order to notify them of blog posts by other developers that fall under their area of interest.

4   Design

The basic features of the IDE are registration and login, creation, saving and deletion of Java projects and files, compilation and execution of these projects.

Our proposed IDE will have a few additional features of real time collaboration, integrated forum and blogging facility. This is shown in Fig 2.

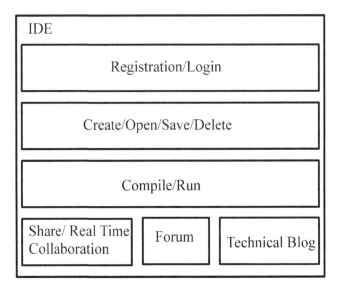

Fig. 2. Architectural blocks in the IDE

The actions of registration and login will be stored in a table. Also, actions of creation, opening and deletion of projects or files and sharing of the projects will also be logged. The details stored in the logs table will include the action performed, the user ID of the developer who performs that action, the time at which the action was performed and if applicable, which file or project the action was performed on.

The three additional important features of real time collaboration, integrated forums and technical blogs that this IDE provides are explained below:

4.1   Real Time Collaboration

Real time collaboration is a technique that allows multiple users to edit the same file at the same time. The edits made by one collaborator will be visible to the other collaborator in real time.

Two tables are required to achieve this- Files table and Access Rights table. This is shown in the Fig 3.

The developer first creates a file. The file ID is stored in the Files table along with the identity of the owner. The developer can then share the file with other developers and users. When the share action is performed, a new entry will be made in the Access Rights table. The new entry will contain the ID of the developer with whom the file is being shared and the ID of the file being shared.

Now, when two or more developers open the same file, a copy of the data within the file will be displayed to all the developers within a text area. These developers can edit the files simultaneously. The edits will be immediately stored in the database. This is explained with the help of Fig 4. The edits that were made will be visible to all the other users simultaneously accessing the file. This will be achieved by real time collaborative editing calls.

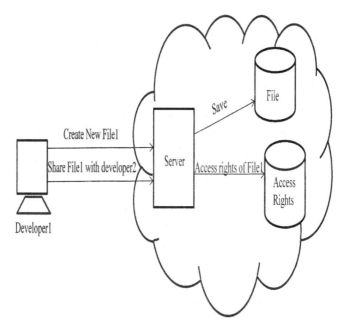

Fig. 3. Creating and sharing files

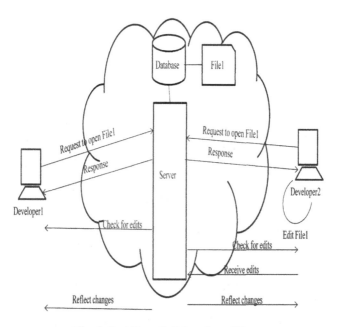

Fig. 4. Real Time Collaborative editing

Thus, through regular real time collaborative editing calls, the edits by certain users can be viewed by other users in real time without having to refresh the page.

4.2   Integrated Forum

A forum is also known as a message board. It is a place where the users can hold discussions, ask questions and answer them. The developers who face certain problems during coding can make use of this in order to get help from other developers. The developers can post questions regarding the error or problems that they are facing and can even post snippets of their code for better clarity of the question to all those who can view it. The other developers who may know how to solve the error or overcome the problem will answer the question.

This IDE will provide an integrated forum that will have access modifiers for the questions that are posted. The developers, who post questions, can select either public or private access modifiers for their posts.

If a developer specifies the access modifier for his post as 'public', then the post will be visible to all the developers using the IDE. Any of the other developers or all of them can post answers to this question. This is shown in Fig 5.

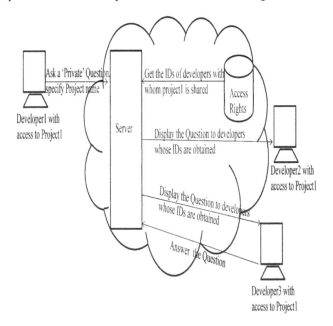

Fig. 5. Posting a 'public' question and answering it

The developer may also want to keep the post as 'private' for security reasons. The developer, in such cases will specify the access modifier as 'private'. He will then mention the name on the project to which the question is related. On doing so, the post by the developer will be accessible to only those developers with whom that project has been shared. This is shown in Fig 6.

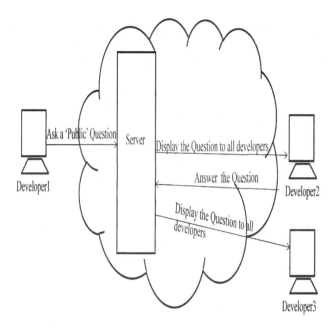

Fig. 6. Posting a 'private' question and answering it

This will be useful when the query is to be asked to only those people who are working on the same project.

The forum will be very useful in situations where the developer wants some help and other developers with whom he has shared the project are currently not online. The developer can post his question and the other developers can answer them later when they access the IDE.

The forum will also award points to the ones who answer questions that are accepted by other developers as 'right'. They will also be given negative marks when the answer is marked 'wrong' by other developers. This point system will encourage more developers to participate and help each other with right answers.

4.3   Technical Blogs

Technical blogs can be used by all the developers using the IDE to post technical information and share it with others. The blog posts will have technical contents related to various topics that may be categorized as 'algorithm', 'swings', 'JSP', etc. The developer who posts information is required to mention the tag i.e. the developer should specify a tag that the information deals with.

The latest blog post title names will be available to all the developers using the IDE. Any of the developers can click on the link in order to view the posts.

Also, a search facility will be provided for the developers to search for the latest blog posts on some topics. For example, a developer may only be interested in seeing blog posts that deal with JSPs. So, he may use the search facility to obtain the blog post titles that have information related to JSPs. This is shown in the Fig 7.

Some developers may not want to see the blog posts or get any notifications of the same. So, they can change their settings such that they will not get alerts regarding the posts. The developers will be allowed to modify these settings anytime according to their choice.

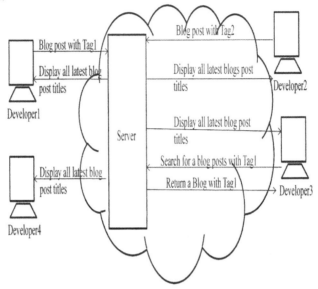

Fig. 7. Blog posts with tags and accessing them

5   Conclusion

Browser based IDE is the need of the hour. There are a few Browser Based IDEs which are primarily for web development. This paper conveys the idea of the creation of a Browser Based IDE to code Java language in the cloud with the additional feature of real time collaboration for the users. It also conveys the idea of integrated forums and technical blog facilities for all the users of the IDE.

References

[1] http://www.techno-pulse.com
[2] http://www.difranco.net/progstuff/voc_list.htm#Integrated_
     Development_Environment
[3] http://www.makeuseof.com/tag/top-3-browser-based-ides-
     code-cloud-2
[4] http://www.wisegeek.com/what-is-an-internet-forum.htm
[5] http://www.blogger.com/tour_start.g
[6] http://cloud9ide.com/
[7] http://coderun.com/

[8] http://wiki.eclipse.org/Orion

[9] http://wiert.wordpress.com/2011/03/18/ideone-com-online-
     ide-debugging-tool-cc-java-php-python-perl-and-40-
     compilers-and-intepreters/

[10] http://www.sdtimes.com/link/35860

Zero Knowledge Password Authentication Protocol

Nivedita Datta

Supercomputer Education & Research Centre
Indian Institute of Science, Bangalore
Bangalore, India
nivedita.mtech@gmail.com

Abstract. In many applications, the password is sent as cleartext to the server to be authenticated thus providing the eavesdropper with opportunity to steal valuable data. This paper presents a simple protocol based on zero knowledge proof by which the user can prove to the authentication server that he has the password without having to send the password to the server as either cleartext or in encrypted format. Thus the user can authenticate himself without having to actually reveal the password to the server. Also, another version of this protocol has been proposed which makes use of public key cryptography thus adding one more level of security to the protocol and enabling mutual authentication between the client & server.

Keywords: computer network, computer security, authentication protocol, zero-knowledge proof, password.

1   Introduction

1.1   Motivation

In today's world of Internet, most of the people have mail accounts or accounts with social networking sites etc. where they need to authenticate themselves before logging in and being able to access their resources. However, very few people are actually aware of the fact that many of such applications make use of PAP(Password Authentication Protocol) in order to authenticate the users which is not very secure.

In case of PAP, though the password is stored in hashed format on the server along with its corresponding username making it less vulnerable to attacks, still the fact that the username-password pair travels in clear on the wire makes it vulnerable to attacks like eavesdropping & packet sniffing which will easily reveal the sensitive data to the intruder.

Here it is assumed that only the user knows the sensitive data(password) which is a secret information for him.

1.2   Contribution

This paper presents a protocol using which the users can be authenticated by the authentication server without having to reveal the password. This protocol, based on

S. Patnaik et al. (Eds.): *New Paradigms in Internet Computing,* AISC 203, pp. 71–79.
DOI: 10.1007/978-3-642-35461-8_7   © Springer-Verlag Berlin Heidelberg 2013

zero knowledge proof[6], thus protects the sensitive data from being revealed to the server or any intruder listening to the communication channel. It is meant to to be basically used in distributed system or peer to peer networks.

This paper first presents a simple version of the ZK-PAP in which the user can authenticate himself to the server without revealing the password[8]. The protocol uses a challenge-response mechanism (between the server and client) based on nonce. A nonce is a randomly generated number to be used only once throughout the session in order to avoid replay attacks.

The simple version of this protocol supports only one way authentication i.e. only the clients can authenticate themselves to the server. However, the other way round authentication is not possible.

The other version of this protocol i.e. ZK-PAP with PKE incorporates the concept of public key cryptography[4] thus adding a second level of security to the protocol and also enabling two-way authentication, i.e. the client can authenticate the server and vice versa.

1.3   Organization of Paper

The paper has been briefly divided into four sections. The first section introduces the readers to the basic notations and concept such as zero-knowledge proof [6,7,10] and PAP [11] which one needs to understand before he can understand the protocol proposed. The second section gives a basic idea about the CHAP authentication protocol which is a relevant work in this area.

The third section gives some brief idea about the basic primitives or building block of the protocol followed by description of working of the protocol proposed in this paper.

2   Notations and Definitions

2.1   Notations

In this section, we shall be discussing some of the basic notations which we will encounter in the paper later. Key $k \in K$ is symmetric session key which will be established between the user and client in every session to carry out the further communication. H is a collision resistant hash function used to generate the hash value of any data. As discussed already, nonce is a randomly generated data denoted by Ni (N1, N2 etc) and transformation function is any simple mathematical function which can be applied on integer data (assuming that nonce here is integer in nature).

Also we have encryption & decryption functions which are denoted by E & D respectively. In case of asymmetric (public key) cryptography, E_{PR-A} & E_{PU-A} represent encryption using private key & public key of A respectively. Similarly, D_{PR-A} & D_{PU-A} represent decryption using private key & public key of A respectively. In case of symmetric (private key) cryptography, as we have no concept of public key hence E_{PR-A} & D_{PR-A} represent encryption and decryption respectively using the secret key of A.

2.2   Definitions

Here we shall be discussing the concepts of zero-knowledge proofs and PAP in brief.

2.2.1   Zero-Knowledge Proof

Let us first discuss the concept of zero knowledge proofs. The concept of zero-knowledge can be explained with the help of a classical example of two identical balls[9]. Suppose a person, say 'A' has two identical billiards ball of different colors, say red and blue. Now he want to convince his friend, say 'B' that the two balls are of different colors.

The basic approach will be to give the two balls to B so that he can see them and confirm the fact that the two balls are of different colors or not. However, in this scheme B gains knowledge about the colors of the balls.

Using the zero-knowledge approach, however A can convince his friend B that he has balls of different colors without having B see the balls actually. To do this, A blindfolds B and then places a ball on each of B's hand. Though B has no idea about which ball is of which color but A can see the color of the two balls.

Now A asks B to take his hands at the back and either swap the arrangement of the two balls or keep the arrangement same as original and show him the balls again. A sees the new arrangement of the balls and lets B know whether the balls were swapped or not. Thus A can prove to B that he has given him balls of different colors without revealing anything about color of the balls.

Let us say they play this game 't' times, where the value of t is large. If A tries to cheat B by giving him both the balls of same color, then the probability that A will still be able to answer correctly in each game is 2^{-t} which is negligible for large value of t.

This is a zero knowledge approach since A convinced B that he has two balls of different colors but at end of all games, B does not gain any knowledge about the colors of the two balls or any knowledge on how to distinguish the two balls.

Another classic example to understand zero-knowledge proof is given in [15] which uses the example of magic cave to explain the same concept.

2.2.2   Password Authentication Protocol

Let us now discuss about the Password Authentication Protocol(PAP)[12,13]. PAP is an authentication protocol which is being used by point-to-point protocol to validate and authenticate users before they can access resources. This protocol requires the user to send the username and password to the authenticating server in cleartext thus making it vulnerable to packet sniffing & eavesdropping.

After the server receives the username & password, it generates hash of the password using the same algorithm which was used to hash the password before storing it into the password file. Then the generated hash is matched against the stored password hash corresponding to the entered user name. If a match is found, then the user is allowed to login else access is denied.

Here, though the password is stored in encrypted format on the server thus making it less vulnerable to attacks but sending the unencrypted ASCII password over the network makes the protocol insecure.

3   Relevance to Prior Work

One of the relevant work done in this field is the CHAP(challenge handshake authentication protocol)[1,2,3,5,12]. This protocol is based on challenge-response model and makes use of single-use keys to provide more security. However this system does not completely eliminate the need to send data over wire in plain text format.

This protocol works in the following manner: when a user types his user name, the server generates a random key and sends it to the client machine(user) in unencrypted format. The user then encrypts his password using the received key and sends it to the server. The authenticator program on server encrypts the password corresponding to the received username using the generated key & matches it against the data received from the client machine.

The user is allowed to login and access his resources if the match occurs else access is denied.

Also, CHAP keeps sending various challenges to the client (user) throughout the session to verify that only an authorized person is logged in.

The main advantages of the scheme are as follows:

◆   It solves the problem of logged in but unattended systems.
◆   Also, the password no more travels in clear but in encrypted form thus solving the problem of packet sniffing or eavesdropping.

However, this scheme poses the following disadvantages :

◆   As the randomly generated key is sent to the user in clear, an intruder can get the key by packet-sniffing.
◆   The password on the server is stored in unencrypted format thus making it more vulnerable to attacks.
◆   Also, on continuously sniffing a line, the intruder will be able collect many key-ciphertext pairs for a user's password thus gaining some knowledge about the user's password.

4   Cryptographic Primitives

The algorithms which are designed to perform any cryptographic operation are known as cryptographic primitives. The primitives are the building blocks which are used to create more complex cryptographic protocols to achieve various security goals. The primitives can be classified into two major groups : symmetric (or private key) & asymmetric (or public key). We will now define some of the primitives used in the proposed protocol:

a) Collision Resistant Hash Function[16]: A collision resistant hash function is a function which takes a variable length input and produces a fixed length output with the property that even slightest change in the input will reflect change in the output(hash value). The input to hash function is called a message and the output is known as hash value.

The collision resistant hash function exhibits the following four properties:

◆ It should be easy to compute hash value for any message.
◆ It should be infeasible to deduce the message from the hash value. This is known as one way property.
◆ It should be infeasible to find two different messages say m1 & m2 with same hash value. This property is known as collision resistance.
◆ It should be infeasible to change a message without reflecting any change in it's hash value.

This kind of hash function has many applications such as digital signature, MAC etc.

b) Block Cipher: It is one of the most important primitives of various cryptographic algorithms & protocols like MAC & various hash functions. It is used mainly to provide confidentiality of data. Block cipher works on fixed length inputs known as blocks. These ciphers encrypt or decrypt one block of data at a time.

Some of the most widely used block ciphers are DES (Data Encryption Standard), 3DES (Triple DES) and AES (Advanced Encryption Standard).

c) Stream Cipher: It is another most important & common primitives for various cryptographic algorithms. In this case, the encryption or decryption of data takes place one bit at a time. Thus it can be treated as a block cipher with block size of 1 bit.

Some of the most commonly used stream cipher algorithms are ORYX, SEAL & RC4.

d) Transformation Function: Transformation function is any simple mathematical function which can be applied to an integer. Here the transformation function is applied on the nonces to avoid replay attacks.

5   The Proposed Security Protocol : Zero Knowledge Password Authentication Protocol (ZK-PAP)

As in general scenario, every user has a username & password used to login to a system to access various resources. The password is secret to the user which only he can change when logged in to the application and the same change is registered with the server.

The simple version of the algorithm provides only one way authentication, that is, only server can authenticate a client system. Let us designate the server and client as verifier and prover for ease of understanding.

The protocol is initiated by the prover by sending his username and a challenge(nonce) N1 to the verifier in clear. The verifier responds by generating a random session key, say k and another challenge(nonce) N2. Then it concatenates N1,

N2 & k and encrypts them using the hash of the password corresponding to the received user name. This encrypted data is then sent to the prover.

The prover now decrypts the data using the hash of its password as key, fetches the values of N1, N2 & K and verifies if the value of N1 received is same as the one it had sent to the verifier. The nonce N1 here is used only to avoid any replay attack. If the value of the received & the generated nonce do not match, then the received message is discarded else it retrieves the session key. The prover then applies the transformation function on the nonce N2, encrypts it with the received session key and sends it to the verifier.

Once the verifier receives the encrypted message, it then decrypts the message with the generated session key and matches it with the expected value. If match occurs, then the user is allowed to login to his account and access resources else access is denied. As in CHAP, in ZK-PAP, a series of challenges can be exchanged between the prover & verifier through out the session to verify that only an authorized person is logged in.

The main advantages of this protocol are as follows:

◆ The authentication is done without the need of the password to travel across the wire.
◆ The password in the password file on server is stored in encrypted format thus making it less vulnerable to attacks.
◆ The security of the protocol mainly depends on the strength of the encryption algorithm being used. Thus using the standard algorithms like AES, DES etc will provide high degree of security to the protocol.
◆ Use of nonce at each step helps us prevent replay attacks.

Here it is assumed that the security of the server is not compromised else the protocol becomes vulnerable to attacks. In this protocol, we can also use time stamp instead of nonce, however that will incur an overhead of keeping all the communicating systems synchronized in time.

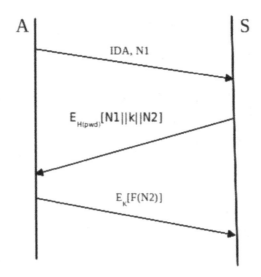

Fig. 1. Zero Knowledge Password Authentication Protocol

Notations used:

IDA: Username of A
N1 & N2: Nonce
k: Shared secret key between A(user) & S(server)
F: Transformation function
E_K**:** Encryption using key k
H[pwd]: Hash of the password

6   ZK-PAP with PKE

This section briefs about the other version of the ZK-PAP protocol described above. This version of the protocol makes use of public key encryption[4] in order to give an added level of security and also enable two-way authentication ie. the verifier(server) can authenticate the prover(client) and vice versa.

Here it is assumed that all the users have (or can get) the public key of the server and the server has or can receive public keys of all the users. The protocol works as follows:

◆   The user, say A sends his username and a nonce to the server after encrypting it with server's public key.
◆   The server decrypts the message with his private key and extracts the value of the nonce N1.
◆   The server then generates a nonce N2 and a random session key k, concatenates N1, k & N2 , encrypts them with hash of the password of user A, then with public key of the user A and sends the encrypted data to A.
◆   User A then decrypts the received encrypted data with his private key, then with the hash of his password and extracts the values of N1, N2 & k. He then matches the value of received nonce N1 & the generated value of N1.
◆   If match occurs, then A extracts the value of k & nonce N2, applies the transformation function F on N2 and encrypts the transformed value first with the session key k, then with public key of the server and sends the encrypted message to the server.
◆   The server decrypts the received value with its private key & then with the shared session key.
◆   The user A is allowed to login if the server receives the expected value else access is denied.

As it can be seen from the above steps, only server will be able to extract the correct value of nonce N1 as it was encrypted with server's public key. Thus, if the client receives correct value of the nonce N1 from the server, it knows that the message was sent by the server itself and not by some intruder. Thus, use of public key encryption also allows the client to authenticate the server thus enabling mutual authentication.

Also, a series of challenge can be exchanged between the server and client to ensure that only an authorized person is logged in. This will also solve the problem of logged-in but unattended systems or workstations.

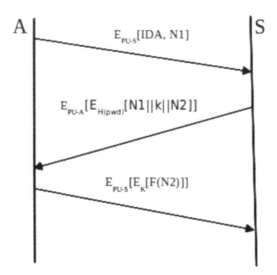

Fig. 2. ZK-PAP with PKE

Notations used:

IDA: User name of A
N1 & N2: Nonce
k: Shared secret key between A(user) & S(server)
F: Transformation function
E_K: Encryption using key k
H[pwd]: Hash of the password
$E_{PU\text{-}S}$ & $E_{PU\text{-}A}$: Encryption using public key of S & A respectively

7   Conclusion

This paper illustrates ZK-PAP and ZK-PAP with PKE protocols, both of which are based on the concept of zero-knowledge proof. The ability to authenticate oneself without having to reveal one's password will make the system less vulnerable to attacks. As the protocol uses the hash of the password as key, using a strong encryption cipher (in which key-recovery is hard) will strengthen the security of this protocol.

Also using the public-key encryption in ZK-PAP with PKE adds a second level of security and enables mutual authentication between the client & server. Both protocol proposed here are simple & efficient, thus enabling their practical use.

Acknowledgment. The author extends thanks to Indian Institute of Science at Bangalore, India for introducing me to this fascinating field in cryptography and giving me the opportunity to study it for my own interest.

References

[1] Simpson, W.: Request for Comments 1994, PPP Challenge Handshake Authentication Protocol (CHAP). Network Working Group, California (1996)

[2] Youssef, M.W., El-Gendy, H.: Securing Authentication of TCP/IP Layer Two by Modifying Challenge-Handshake Authentication Protocol. Advanced Computing: An International Journal (ACIJ) 3(2) (March 2012)

[3] Zorn, G.: Request for Comments: 2759: Microsoft PPP CHAP Extensions- Version 2, Network Working Group, Microsoft Corporation (2000)

[4] Dolev, Yao, A.: On the Security of Public Key Protocols. IEEE Transactions on Information Theory 29(2), 198–208 (1983)

[5] Verification of two versions of the Challenge Handshake Authentication Protocol (CHAP), Guy Leduc, Research Unit in Networking (RUN)

[6] Goldreich, O.: Zero-knowledge twenty years after its invention (2002) (unpublished manuscript)

[7] Zero-knowledge proof. Wikipedia, The Free Encyclopedia, http://en.wikipedia.org/wiki/Zero-knowledge_proof

[8] Zero-knowledge password proof Wikepedia, The Free Encyclopedia, http://en.wikipedia.org/wiki/Zero-knowledge_password_proof

[9] Challenging epistemology: Interactive proofs and zero knowledge Justin Bledin Group in Logic and the Methodology of Science. Journal of Applied Logic 6, 490–501 (2008)

[10] Mohr, A.: A Survey of Zero-Knowledge Proofs with Applications to Cryptography. Southern Illinois University, Carbondale

[11] "Password Authentication Protocol" Wikipedia, the free encyclopedia, http://en.wikipedia.org/wiki/Password_authentication_protocol

[12] Microsoft TechNet, Authentication Methods, http://technet.microsoft.com/en-us/library/cc958013.aspx

[13] Microsoft Technet, Password Authentication Protocol, http://technet.microsoft.com/enus/library/cc737807%28v=ws.10%29

[14] Lloyd, B., Simpson, W.: Request for Comments 1334. PPP Authentication Protocols, Network Working Group (October 1992)

[15] Quisquater, J.-J., Guillou, L.C., Berson, T.A.: How to Explain Zero-Knowledge Protocols to Your Children. In: Brassard, G. (ed.) CRYPTO 1989. LNCS, vol. 435, pp. 628–631. Springer, Heidelberg (1990), http://www.cs.wisc.edu/~mkowalcz/628.pdf

[16] Cryptographic Hash Function. Wikipedia, the free encyclopedia, http://en.wikipedia.org/wiki/Cryptographic_hash_function

Author Index